Universal Kinship

The Bond Between All Living Things

Edited by
The Latham Foundation Staff

The Latham Foundation
Clement & Schiller
Alameda, California 94501

Published by
R&E PUBLISHERS
P.O. Box 2008
Saratoga, California 95070
(408) 886-6303

Library of Congress Cataloging-in-Publication Data
L.C. 91-50981

ISBN 0-88247-918-0
Copyright © 1991
The Latham Foundation

All rights reserved. No part of this book may be reproduced, stored in a retrieval system, or transmitted in any form, or by any means, electronic, mechanical, photocopying, recording or otherwise without prior permission of the copyright owner.

Cover illustration by Chung Ngatpor, winner of the Latham International poster contest, 1952, Westlake Hangchow, China, National Institute of the Arts

Cover design and text illustrations by Kaye Quinn, winner in the Latham International poster contest, 1957

Typesetting by elletro Productions

PREFACE

As the end of the 20th Century approaches, we have begun to reach a growing awareness of the fragility of our living planet. The technological revolution, with all its wonders, has created a new tyranny of profligate usage of limited natural resources, overwhelming amounts of poisons that threaten all life, stockpiles of nuclear, chemical, and biological weapons of mass destruction and increased isolation from nature, humanity and our own families.

It is time for us to go full circle, to return to an ancient, perhaps primeval, wisdom that understands the interrelation and interdependence of all living things and the need for harmony and cooperation. It is time that we realize that we are caretakers, not owners of the planet upon which we live.

In 1918, the Latham Foundation was created to promote human awareness of the importance to its very existence, for it to respect the life and well being of all our fellow traveling creatures on this planet.

THE LATHAM LETTER has, over the years, served to publicize the inter-disciplinary aspects of human companion animal relationships and other vital social and ecological matters. Its studies have rationally explored those important subjects from scientific as well as applied standpoints.

"UNIVERSAL KINSHIP: The Bond Between All Living Things" preserves in this more permanent form, a selection of THE LATHAM LETTER's outstanding articles. It is its purpose to provide a summary of subjects with essential information in brief form, for the student or applies practitioner, and to serve as an encouraging guide for future studies and investigation. The Latham Foundation is particularly grateful to the nationally and internationally respected authorities concerned, for permission to reprint their excellent work in this volume.

Hugh H. Tebault

President - The Latham Foundation

Acknowledgements

This book could not have been published, nor its readers benefitted, without the cooperation of the outstanding and compassionate thinkers whose work is recorded within its covers. To them, The Latham Foundation extends its sincere and grateful appreciation.

The title "Universal Kinship" and the graphic appearing on the book's cover, were created by Chung Ngat Por, of the National Institute of the Arts, Hangchow, China, winner of the 1952 Latham Foundation Poster Contest.

Contents

1. HCAB/PFT 11

Special Needs for the Pet Owner with AIDS/HIV
Ken Gorczyca, DVM 13

Effects of Animal-Assisted Therapy on Communications Patterns with Chronic Schizophrenics
*Lawrence Bauman, Monte Posner,
Karl Sachs, and Robert Szita* 21

Effects of Pets on the Well Being of the Elderly
*Cathleen M. Connell, M.S.
Daniel J. Lago, Ph. D.* 51

Animal Assisted Therapy in the Psychiatric Setting
Christine Shaheen, AAT 63

Stressful Life Effects and Use of Physician Services Among the Elderly: The Modifying Role of Pet Ownership
Judith M. Siegel, Ph. D. 71

2. GRIEF 91

When the Bond is Broken: Companion Animal Death and Adult Human Grief
Mary M. Bloom, M.A. 93

Death With Dignity
*Thomas E Catanzaro, D.V.M.
Jodie L. Sell* 101

Pet Loss Considered from the Veterinary Perspective
Eddie Garcia, D.V.M. 119

Grief Counseling for Euthanasia
Cecelia J. Soares, D.V.M., M.S. 133

3. CHILDREN 143

Therapy Dog in the Classroom
Dee Press, M.A. 145

A Time of Innocence: A Time of Violence
Marcia Kelly 157

Reaction of Infants and Toddlers to Live and Toy Animals
Aline H. Kidd, Ph.D.
Robert M. Kidd, M.A., M.Div. 167

Pets and the Socialization of Children
Michael Robin
Robert ten Bensel M.D. 173

4. THE ENVIRONMENT 197

Population Growth and the Environment
.... 199

Environmental Protection for the 1990's and Beyond
Milton Russell 205

5. PHILOSOPHY 227

The Evolution of Animals in Moral Philosophy
Steve F. Sapontzis, Ph. D. 229

Living Miracles
The Reverend Dr. Andrew Linzey 239

6. MISCELLANEOUS 247

Unexpected Teachers: A New Look at our Pets
Karen Kaufman Milstein, Ph. D. 249

Webs
Stephan H. Johnsrud 257

HCAB\PFT

Human-Companion Animal Bond
Pet Facilitated Therapy

Special Needs for the Pet Owner With AIDS/HIV

PETS ARE WONDERFUL SUPPORT

Ken Gorczyca, DVM

Acquired immune deficiency syndrome (AIDS) is a growing problem of epidemic proportions in the United States and the world. Those who have not already done so can expect to have some contact with a person infected with the human immunodeficiency virus (HIVer). This virus is primarily transmitted through exposure to infected human blood components, and prenatally from mother to child. Studies of HIVers indicate that infection is not transmitted through casual contact with saliva or tears. Unfounded fears have many times isolated HIVers from their family and friends. HIVers often undergo emotional, physical and financial difficulties. The loss of friends, family, lovers and employment often leads to emotional distress. The loss of health often leads to physical disabilities, loss of energy and loss of mobility. The high cost of medical care and the loss of employment many times leads to a substandard of living. Thus, companionship, both human and animal, can play an important role for HIVers.

Recent studies report that animal-assisted therapy has allowed for dramatic and rapid positive physiological and psychological changes in the elderly and disabled. In one study, cardiac patients with pets lived longer than cardiac patients without pets. Most recently, another study showed that senior citizens with dogs had fewer doctor visits over a one-year period than seniors without dogs. Research on the healthy population has shown that blood pressure drops when a person strokes a dog or watches a fish. Animal companions are continuing to be shown to be of much value for all of us. HIVers who may feel isolated, rejected, and stigmatized by other people often find continuous, non-judgmental love in their animal companion. The quality of life is surely improved and, possibly, the longevity as well.

Since the beginning of the AIDS epidemic in 1981, many organizations have been developed to help HIVers. The San Francisco AIDS Foundation helps provide up-to-date information about AIDS and AIDS related issues. The Shanti Project helps people with AIDS (referred to as PWAs) directly through emotional and practical support. Project Open Hand helps feed PWAs. In San Francisco alone, there are over 100 organizations that assist HIVers with a variety of services. However, initially there was very little support for pet owners with AIDS/HIV. Those people who chose to keep their pets often found that financial or physical constraints made that choice unrealistic. Existing animal-oriented organizations were not set up to provide the kind of in-home support that these people needed to be able to keep their pets at home with them.

In 1987, a group of people banded together to do whatever was necessary to help their friends with AIDS/HIV keep their pets for as long as possible. In recognition of the human-companion animal bond, they called their group **Pets Are Wonderful Support**, nicknamed **PAWS**. This organization has sought to fill in the gaps between other AIDS services and animal related organizations and address the particular problems and questions faced by the immunosuppressed pet owner.

Special Needs For The Pet Owner With AIDS/HIV

Initially, PAWS helped supply pet foods to the AIDS Food Bank. Over time, it became apparent to the group of volunteers that there were many other needs for the pet owner with AIDS/HIV. The financial needs of pet care were obvious—veterinary care and pet food. The emotional needs were more discreet. What happens to the animal if the human companion suddenly goes into the hospital? What if the person becomes too weak to walk their dog or change the aquarium? What happens to someone's precious companion if the person dies? What are the risks, for humans, of diseases that can be acquired from animals, especially if they are immuno-compromised? What about the worry caused by all these uncertainties?

Pets Are Wonderful Support exists to improve the quality of life for persons with HIV (AIDS/ARC) by offering them emotional and practical support to keep the love and companionship of their pet(s) and by providing information on the benefits and risks of animal companionship.

In early 1987, PAWS had twelve clients. Today, there are over 400 pet owners with AIDS/HIV who are assisted with a variety of services. The volunteer force has steadily grown from a few steady, motivated individuals to over 50 active volunteers. The growth has been exponential and more dedicated volunteers are needed to help run the many programs.

In November, 1988, PAWS hired its first part-time employee to help run the office. By the fall of 1989, we hired an executive director, Leah Talley, who has been instrumental in gearing up volunteers, helping clients and developing community education. The human-animal care programs presently include foster care, adoption, in-home pet care, pet foods, veterinary care and community education.

PAWS' foster care program not only helps the animal, it helps the HIVer emotionally. PAWS offers short-term homes for pets of clients who need a break or are temporarily unable to care for them. Many times, a person can become ill suddenly and have to "abandon" their pet on short notice. We try to match the animal with a home from a list of volunteers who are able to offer a temporary home. This program is ideal for volunteers who enjoy the company of animals, but who are unable to keep them, on a full-time basis.

The in-home pet care service involves anything from dog walking or litter box changing to aquarium cleaning. The PAWS volunteers are paired up with a client. By using the buddy system, the volunteer, client and companion animal are able to develop a caring, secure and comfortable relationship. This extra help allows many pet owners to keep their friend much longer. There are many AIDS services that help feed, provide financial assistance, and provide medical information. PAWS helps people and pets.

The PAWS pet food bank is our most popular service. We offer home delivery of pet foods and supplies to our clients who cannot leave their homes. We hope to supplement this service by continued donations of pet foods during the 1989 California Veterinary Medical Association Conference. We welcome donations of dog and cat foods.

Veterinary care has become our biggest expenditure. We help pay for veterinary care for clients who have a proven financial need. We utilize veterinary hospitals which offers transportation to and from the clinics.

Our adoption service is one of our most important functions. PAWS helps find good homes for pets of clients who have died of AIDS, or for animals that can no longer be kept by the HIVer. In most cases, if a person is moved to a hospice or nursing home, they are not allowed to keep their pet. We try to make this loss easier by helping to find a suitable adoptive home for their pet. We also try to provide the opportunity for the person giving up their pet to meet the new family. We have been very successful in adopting out older pets.

The benefits of pets for people are quite obvious, but the risks of pet ownership for the immunosuppressed person are not. Such a person is prone to developing many infections that a healthy person can normally fight off. The list of infections includes those that can be acquired from animals, or zoonoses. Although the risk of zoonoses is small, there is a lot of controversy surrounding this issue. The medical community was not prepared to answer questions about zoonoses. Veterinarians know about animal-borne diseases, but little about the increased risk to immunosuppressed humans. Prudent physicians, unfamiliar with the details of zoonotic disease transmission, often chose to err on the side of caution, and simply advised their patients with AIDS to get rid of their pets. Ironically, they were the very people who most needed the emotional and psychological benefits their animal companions gave them. Luckily, these attitudes have been changing and most physicians support animal companionship for their patients with AIDS.

Pets Are Wonderful Support started to educate the medical, veterinary and AIDS communities about zoonoses, AIDS and the human-companion animal bond and safe pet care. The organization published the first edition of the brochure, Safe Pet Guidelines, in 1988.

Pets can be made relatively more safe by following PAWS guidelines. Common sense, controlling the pets' diet and environment, good hygiene (soap and water!), and keeping the pet healthy all lead to a decreased risk of disease transmission from pet to person. Our latest edition of Safe Pet Guidelines has been generalized to include other groups of immunocompromised persons which include people on chemotherapy, those having certain types of cancers, pregnant women and the aged. We have distributed over 15,000 brochures worldwide. Pets Are Wonderful Support is evolving into a national clearinghouse for information on zoonotic diseases, immunosuppressed individuals and animal companionship. PAWS has presented talks and papers at numerous medical, veterinary, AIDS, human-animal bond and gay conferences. Our active education committee meets quarterly to help develop our educational needs. PAWS has received grants to produce a brochure on the controversial disease, Toxoplasmosis, from the Bay Area Physicians For Human Rights Foundation. We also received a grant from the Horizons Foundation to set up and run a Zoonoses Advisory Board made up of physicians, veterinarians and other professionals to help guide the PAWS educational program.

Pet owners with AIDS/HIV have specific emotional, practical and educational needs. Animal companions contribute to the quality of life and possibly the longevity, but also hold some risks which must be addressed. Pets Are Wonderful Support is trying to address these needs.

Dr. Gorczyca is the Director of Education for Pets Are Wonderful Support. He is also a practicing veterinarian. For more information about PAWS or its brochure, Safe Pet Guidelines, contact Pets Are Wonderful Support at P.O. Box 460489, San Francisco, CA 94146.

The Latham Letter, Vol. XI, No. 4, Fall 1990, pp. 1, 18-19.

The Effects of Animal-Assisted Therapy on Communication Patterns With Chronic Schizophrenics

Lawrence Bauman, Monte Posner, Karl Sachs, and Robert Szita

INTRODUCTION

For the past 15 years, companion animals have been regularly used in an adjunctive capacity for treating physically and mentally ill people. Proximity to a variety of pets (cats, dogs, kittens, puppies, fish, and parakeets) has been associated with increased survival rates among coronary heart diseased patients (Friedmann, Katcher, Lynch, & Thomas, 1980), as well as potentially life-saving improvements in blood pressure among the institutionalized elderly (Daniel, Burke, Rutel, and Burke, 1987; Katcher, 1982).

Similarly, pet ownership was observed to mitigate the effects of stressful life events by contributing to fewer doctor contacts among a large group of elderly survey respondents (Siegel, 1990). The latter study points out the potential importance that pet companions have in soothing the spirits of fortunate pet owners.

To date, the body of work substantiating the psychological benefits of human/companion animal interactions is steadily expanding. Mere visual exposure to puppies prompted elderly people in long-term care to demonstrate increased rates of verbalization, smiling and wide-open eyes (Robb, Boyd, & Pritash, 1980). Among a group of noninstitutionalized old people, exposure to caged parakeets fostered improved morale (Mugford & M'Comisky, 1975).

When domesticated cats and dogs were introduced in institutional environments, the impact on social behavior increased in seeming proportion to the level of exposure. In one study (McCulloch, 1981), outpatients who were depressed, secondary to physical illness, became more spirited when exposed to pet animals. Similarly, at a hospital-based intermediate care unit, middle-aged and elderly OBS patients who were exposed to cat mascots were assessed to be more responsive, more in touch with reality, and friendlier to other patients than before the introduction of cats (Brickel, 1979). At a nursing home for terminally ill patients, allowing residents to hold, caress, talk to, play with and watch kittens and puppies led to projected feelings of love, demonstrative caring, increased pleasure, as well as reduced fear, despair, loneliness, and helplessness (Muschel, 1984).

Despite the proven efficacy of using companion animals with sub-populations as diverse as the medically ill, the elderly, wheel-chair bound children and imprisoned felons, only a few studies have documented the benefits of animal-assisted therapy with psychiatric patients.

Among the exceptions are studies which highlight the role played by animals in promoting adaptive behaviors in institutionalized patients. In a landmark study, a husband and wife team of researchers (Corson & Corson, 1982) observed an increase in verbalizations, emotional expressiveness and reduced tension level in a group of elderly psychiatrically disturbed patients who were formerly unresponsive to traditional modes of treatment. Earlier, this same team of investigators reported decreased question/answer intervals, increased number if questions asked, and increased words per answer with psychiatric patients who heretofore were withdrawn, uncommunicative, or mute (Corson & Corson and Gwynne, 1975).

More recently, when emotionally disturbed children (age 5-17) in residential care were allowed to interact with dogs, the incidence of acting out and aggression was significantly reduced (Daniel, Burke, Comprecht, & McLaren, 1987).

Although companion animals have been introduced with psychiatric patients few, if any, studies have utilized animal-assisted therapy with chronic schizophrenic individuals in residential care. Furthermore, use of animals with institutionalized populations has been informal and often lacking the methodological precision of other therapies or other research endeavors. Since schizophrenic symptomatology is often characterized by aberrations in the quantity of verbalizations (hyperverbal or hypoverbal speech), quality of verbalizations (relevant or irrelevant speech) as well as other social behavior indicators (e.g., erratic eye contact, bizarre body movements, etc.) this team of investigators wanted to collect observational data associated with these variables under two conditions: animal present and animal absent. The primary purpose of the study was to determine if psychiatric patients who were exposed to a variety of animals during bi-weekly group therapy meetings would demonstrate increases in appropriate speech and other social behaviors as compared with a group exposed to a curriculum of mental health concepts.

PROCEDURES

Program

The Community Residential Treatment Service (CRTS) Continuing Treatment Program of South Beach Psychiatric Center provides treatment services for clients with psychiatric problems under the auspices of the N.Y.S. Office of Mental Health. The CRTS Continuing Treatment Program has been in existence for the past 10 years. The primary purpose of the program is to provide aftercare and, in most cases, day programming for psychiatric patients. These patients characteristically have had multiple hospitalizations and have been difficult to place and stabilize in any outpatient setting. The patients reside on the grounds of the hospital in a variety of structured residences, "eighthway," "quarterway," and "halfway houses."

The CRTS Continuing Treatment Program is a highly structured behaviorally oriented, token economy program which provides training in skills necessary for community living, e.g., knowledge of and compliance with medication, appropriate communication skills, stress management, and problem solving.

The clients range in age from 21 to 66. They come from a variety of ethnic, religious, social, and economic groups. These clients have chronic psychiatric histories and a history of unsuccessful attempts at community placements.

The CRTS Continuing Treatment Program is staffed by an inter-disciplinary treatment team. The major focus of the unit is to help the clients through the use of psychotropic medication, behaviorally-oriented skills training groups, individual therapy and family treatment when indicated.

SAMPLE CHARACTERISTICS

Participants in the Animal-Assisted Therapy Project were 19 mostly chronic schizophrenic, single men and women ranging in age from 25 - 69. The mean age was 37 and the average number of previous psychiatric hospitalizations was 4.75. A more comprehensive description of sample characteristics broken down according to group (experimental or control) is presented in Table I.

Table I

Characteristics of Study Participants

Experimental (A.A.T.)

Characteristics	Group	Control Group
Number	10	9
Sex		
Male	80%	67%
Female	20%	33%
Diagnosis		
Schizophrenia	60%	78%
Affective Disorder	0%	11%
Other	40%	11%
Mean # Psychiatric Hospitalizations ('82-'89)	3.6	5.9
Mean Length of Stay Per Hospitalization (in months)	2.8	2.2

PROCEDURES

The CRTS Day Program Animal-Assisted Therapy Project was designed as a three-phase study. Prior to phase 1, the staff of the program (1 psychologist, 1 social worker, 1 registered nurse, and 1 mental health therapy aide) independently rated each of the study participants according to the average frequency of displayed eye contact (E.C.), relevant conversation (R.C.), appropriate body posture (B.O.), and overall participation (F.P.) during routine problem-solving therapy groups. After calculating each patient's current level of social functioning (a composite score of E.C., R.C., B.O., and F.P. ratings), patients' names were then randomly assigned to the experimental (animal-assisted therapy) or control (psychoeducational concepts curriculum) group. In order to ensure initial parity between groups, a T-test was performed using a comparison of grand means and found to be nonsignificant (t=.27, d.f.=17, p=.05).

All participants in the study were volunteers from a total of 19 patients attending the CRTS Continuing Treatment Program and living on the grounds of South Beach Psychiatric Center. In order to protect patients' rights and welfare, each prospective research participant met with his/her therapist or the principal investigator who informed him of: (1) the approximate starting and ending date of the project, (2) the voluntary nature of his/her participation, and (3) the right to withdraw from the project at any point.

EXPERIMENTAL GROUP PROCEDURES

(animal assisted)

Patients assigned to the experimental group received four 45-minute sessions (2/week) of a psychoeducational curriculum entitled, "Building Mental Health" during the baseline phase, followed by six 45-minute (1/week) of animal-assisted therapy (treatment phase), followed by four 45-minute sessions (2/week) of the psychoeducational curriculum [post-treatment phase]. Baseline and post-treatment groups were conducted by a regular CRTS Day Program staff member who also distributed token economy supplies exchangeable for edible or nonedible reinforcers in accordance with the usual day program procedure. For the duration of the research project, the same standard for the reinforcement of "appropriate behavior" was applied to both groups. During the baseline, treatment, and post-treatment phases, two staff members (other than the group leader) recorded the frequency of targeted behaviors at five-minute intervals. A "blind" observer, who was not familiar with the research design, recorded in narrative form, also at five-minute intervals, information pertaining to attentiveness, alertness, and relatedness to the speaker and/or the animal.

Animal-Assisted Therapy (A.A.T.) Sessions

A. PROGRAM OVERVIEW

Each of the six A.A.T. sessions began with a greeting by Mr. S., Program Coordinator for Special Audiences at the Staten Island Zoo. After introducing himself, Mr. S. asked each patient to say his name or addressed by name the patients he already knew. He then informed patients that he would be visiting weekly for six sessions and would bring a different animal each time.

The order of presentation was as follows:

Week	Animal
1	rabbit
2	rooster
3	ferrets (weasel family)
4	goat
5	turtles
6	rabbit (different from week 1)

After the first session, introductory comments reflected where in the six session course the current session belonged (for example, beginning the third session he said, "This is our third meeting ..."). A brief statement pertaining to A.A.T. followed: "We will be learning about some zoo animals and how to get along with them... We will also learn more about ourselves."

B. Animal Introduction

Prior to taking the demonstration animal out of its enclosed carrying case, Mr.S. attempted to desensitize the group from feelings of apprehension or anxiety by explaining how the animal came to the zoo (e.g., born there, dropped off there) and providing other background information (e.g., "This rabbit is friendly. It has been petted by many children at many neighborhood schools.") Next, Mr. S. introduced the animal to the group (e.g., "This is a goat; his/her name is _____."

C. EXPECTATIONS

The group was instructed as to proper voice tone and petting motion for interacting with the animal.

D. INVITATION TO PET OR TOUCH THE ANIMAL

Patients were invited to touch/pet the animal one at a time. Reluctant patients were encouraged to get a closer look at the animal but nobody was forced to come closer than was comfortable. If a patient became anxious or agitated by the mere presence of an animal, he/she was escorted from the group and received supportive counseling.

E. MONITORING

While patients were petting the animal or observing others petting the animal, Mr. S. invited other group members to report what happened as the two interacted. He asked open-ended questions such as, "What happened when T.R. started

to pet the rabbit?" Further questions focused on feelings attributed to the animal, e.g., "How do you think the rooster feels when T.R. touches its neck? Do you think the rabbit is comfortable?"

F. GROUP PROCESS

Throughout the group, Mr. S. solicited questions and comments reflecting the group members' personal feelings, e.g., "How would you feel if you lived in a cage? Do you ever feel lonely living at SBPC? Have you ever felt like that -- carefree or free?"

CONTROL GROUP PROCEDURES

Each patient assigned to the control group received four 45-minute sessions (2/week) of a psychoeducational curriculum entitled, "Building Mental Health," followed by six 45-minute sessions (1/week) of a different section of the same curriculum, followed by four 45-minute sessions (2/week) of the original curriculum. The first and last four sessions constituted the pre and post-treatment phases while the middle six sessions constituted the comparative phase. As with the treatment group all sessions were conducted by a regular CRTS staff member who distributed token economy supplies in accordance with the usual day program procedure. In addition, two staff and one blind observer recorded pertinent data at five-minute intervals.

Following the post-treatment phase, all patients in the study received six sessions conducted by Mr.S., Program Coordinator for Special Audiences at the Staten Island Zoo.

RESULTS

Although T-test comparison of the experimental and control groups prior to beginning the baseline phase failed to reveal initial group differences T-tests performed after completion of the baseline period but prior to the treatment phase revealed some differences. The variables, eye-contact (E.C.)

and irrelevant verbalizations (I.R.) failed to discriminate between the groups, thereby reflecting initial parity, whereas the A.A.T. group had better body orientation scores but fewer relevant verbalizations during the baseline period. Table II provides a summary description of T-test comparisons during the baseline phase.

Table II

Baseline Phase Comparisons
Variables

	Eye Contact		Body Orientation		Relevant Verbalization		Irrelevant Verbalizations	
A.A.T. Control	A.A.T.	Control	A.A.T.	Control	A.A.T.	Control	A.A.T.	Control
Means	5.02	4.29	4.94	3.39	3.38	5.73	1.07	3.26
T-Test Value	1.5		2.2		3.45		1.73	
d.f.	17		17		17		17	
Probability Level	.05		.05		.05		.05	
Statistical Significance	Nonsignificant		Significant		Significant		Nonsignificant	

Since only the variables of eye contact and irrelevant verbalizations failed to discriminate between the groups prior to introducing the respective treatment interventions, the statistical comparisons employed were primarily within-group comparisons. Between-group comparisons were made for eye contact and irrelevant verbalizations since the groups were initially matched on these variable. Table III summarizes the data used to make statistical comparisons. The entries which appear in rows "E.C." and "B.O." refer to the average number of times each of these behaviors occurred during a 45-minute session when observations were recorded at five-minute intervals only. By contrast, rows "R.V." and "I.V." refer to the average number of times each of these behaviors occurred over the course of a 45-minute session when there was no such restriction on recording.

Table III

Comparison of Communication Variable by Group and Across Conditions:

	Dependent Variables	Pre-Treatment (Baseline)	Treatment	Post-Treatment
Experimental Group	1 E.C.	5.02	6.96	5.21
	2 B.O.	4.94	6.87	5.36
	3 R.V.	3.375	9.02	9.625
	4 I.V.	1.07	.174	1.055
Control Group	1 E.C.	4.29	5.3	4.76
	2 B.O.	3.39	4.87	4.38
	3 R.V.	5.73	7.98	7.70
	4 I.V.	3.26	2.35	.66

1 Eye Contact

2 Body Orientation

3 Relevant Verbalizations

4 Irrelevant Verbalizations

EYE CONTACT

Between group comparison of means for eye contact revealed significantly more appropriate eye contacts for the A.A.T. group versus the control group (6.96 vs. 5.3, t = 2.37, d.f. = 16, p = .05) at the completion of the treatment phase.

Within group comparisons further substantiate this finding because the magnitude of increase for eye contact from baseline to treatment phases was significant for A.A.T. patients but not for the controls. Similarly, when A.A.T. was withdrawn in the post-treatment phase, the frequency of eye contact was significantly reduced whereas for the control group no such reduction occurred. Table IV summarizes the impact of the respective treatments on eye contact across conditions.

Table IV

T-Test Comparison of Means for Eye Contact by Group and Across Conditions

Group	Baseline Treatment	Treatment Post-Treatment
A.A.T.	Significant (t=2.62, d.f.=16, p=.02)	Significant (t=2.26, d.f.=13, p=.05)
Control	N.S. (t=1.61, d.f.=16, p=.05)	N.S. (t=.76, d.f.=16, p=.05)

BODY ORIENTATION
(Posture)

Within the group comparisons for the variable body orientation (B.O.) indicate the differential impact of the respective treatments. Patients in the A.A.T. group demonstrated a significant improvement in posture while exposed to the animals (t=3.25, d.f.=17, p=.01), whereas control group members did not (t=1.77, d.f.=16, p=.05 is nonsignificant). Similarly, withdrawal of A.A.T. during the post-treatment phase produced a significant decline in posture (t=2.19, d.f.=13, p=.05), while control group participants experienced no such change following withdrawal of the mental health concepts curriculum (t=.653, d.f.=16, p=.05 is nonsignificant).

RELEVANT VERBALIZATIONS

Unlike variables eye contact and body orientation, the impact of A.A.T. and the alternate treatment on verbal relevance was quite similar. In fact, the "relevant verbalizations" variable was most dramatically effected by both treatment interventions. For both groups, introduction of a novel treatment resulted in significant and sizeable increases in relevant speech. (A.A.T.: t=3.98, d.f.=16, p=.001; Control: t=2.39, d.f.=16, p=.05). Table V illustrates the magnitude of change increase slightly during the post-treatment phase for the A.A.T. group, while a small decline from mid-phase was evident among control group participants. Nevertheless, these subtle changes were nonsignificant following withdrawal of the respective treatments (A.A.T.: t=.32, d.f.=13, p=.05; Control: t=.26, d.f.=16, p=.05). Therefore, in both groups the rate of relevant verbalizing remained at a high level after withdrawal of the respective interventions.

Table V

Percentage Change for Communication Variables Across Conditions

	Dependent Variables	Baseline Treatment	Treatment Post-Treatment	Baseline Post-Treatment
Experimental Group	1 R.V	+268%	+12%	+280%
	2 I.V	——	——	——
Control Group	1 R.V	+139%	-6%	+133%
	2 I.V	-28%	-72%	-80%

1 Relevant Verbalizations

2 Irrelevant Verbalizations

IRRELEVANT VERBALIZATIONS

Between group comparison of means for the variable "irrelevant verbalizations" at completion of the treatment phase revealed significantly fewer irrelevant statements for A.A.T. participants as compared with the control group (.174 vs 2.35, t=4.62, d.f.=16, p=.001). This partially substantiated by the finding of a significant, although not particularly meaningful, decline in irrelevant statements following introduction of A.A.T. (1.07 vs. 174, t=2.42, d.f.=16, p=.05). In contrast, no such decline occurred for the control group (3.26 vs. 2.35) when the "novel" curriculum was introduced (t=.65, d.f.=16, p=.05 is nonsignificant). Surprisingly, the number of irrelevant statements declined in a significant direction (2.35 vs. .66, t=3.18, d.f.=16, p=.01) only *after* the alternate curriculum was withdrawn.

DISCUSSION

Historically, efforts to document the beneficial effects of using animals as adjunctive therapeutic agents with institutionalized, as well as noninstitutionalized populations have been highly anecdotal or based upon consensually-validated clinical observations. Yet, these studies have laid an important groundwork for understanding the depth and variability of human responses to the presence of domesticated cats, dogs, fish, birds, and others. Interacting with animals has produced meaningful and occasionally life-span promoting changes in verbal, non-verbal, and physiological measures. The present study adds another dimension to the earlier investigations since it documents improvements in social behavior with small groups of institutionalized mentally ill patients using controlled, experimental procedures.

The decision to use zoo animal in an adjunctive treatment capacity with chronic schizophrenic patients was based on two factors: (1) Proximity to the Staten Island Zoo and willingness of its zoo coordinator, Mr. S. to run A.A.T. groups and, (2) ongoing use of behavioral treatment technologies (i.e., token economy approaches) to teach appropriate social skills to this population at the Community Residential Treatment Service. Since schizophrenia is partially characterized by aberrations in verbal and nonverbal behaviors (i.e., disconnected and illogical speech, bizarre or inappropriate facial and bodily gestures), we wanted to see if the presence of animals in a structured therapy format would mitigate against this symptomatology.

As demonstrated, animal assisted therapy had a very positive effect on all of the measured variable with the possible exception of irrelevant speech. In both groups, few irrelevant remarks were observed regardless of treatment condition. Therefore, drawing conclusions about the efficacy of A.A.T. as related to irrelevant speech is at best suggestive of a trend.

Despite this exception, A.A.T. produced significant increases in eye contact and appropriate posture. The alternate curriculum (mental health concepts discussion group) did not. Equally telling is the finding that withdrawal of A.A.T. produced a significant decline in the E.C. and B.O. Regression to baseline levels attests to the fact that patients missed the animals when they were gone. Watching and interacting with the zoo animals therefore facilitated more visual focusing and fewer extraneous bodily movements.

The "verbal relevancy" variable was most dramatically affected by A.A.T. as well as the alternate treatment. For both groups, introduction of a novel treatment accounted to an average increase of from 139% - 268% from baseline (See Table V) depending upon condition. For both groups, it appears that meaningful or context-dependent verbal behavior is similar to the phenomenon of a dam barrier, it becomes difficult to slow down, let alone stop to any degree. In our study, speaking in a relevant manner seemed to "take on a life of its own" even after the respective treatments (A.A.T. or the mental health concepts discussion group) were withdrawn. Perhaps, as our patients became interested enough in the animals or concepts that had immediate relevance to them (i.e., how to remain optimistic, to fulfill responsibilities, to keep problems in perspective, etc.), there was little room for the kind of idiosyncratic self-preoccupation that they usually engage in.

Irrelevant speech, by contrast, seems to follow a different set of rules. Few irrelevant statements were observed among experimental (A.A.T.) group participants. Among control group participants, although only 3 irrelevant statements occurred on average during each 45-minute session (baseline phase), the number of irrelevancies steadily declined from mid-point through the post-treatment phases (see Table III). Perhaps irrelevant verbalizations are inversely related to relevant statements. As schizophrenic patients become more relevant and organized in thinking and consequently more expressive in their speech, there is less room or need for irrelevant speech.

It is not clear from our study if the negative relationships between relevant and irrelevant components of speech is a valid finding or if it is an artifact of unintentional rater bias. Staff who collected data are all seasoned clinicians who may have inadvertently paid more attention when relevant statements were made since integrated, socially related speech is a rarer phenomenon for our patients. If relevant speech was disproportionately affected by rater bias, it may be because therapist rates were wishing that the patients could behave in more normal ways just like one hopes a new medication will produce desired effects for a serious physical illness. This hypothesis is only conjecture and was not shown to occur.

In conclusion, the use of animal-assisted therapy in directed small discussion groups with chronic schizophrenic patients was successful in improving selected aspects of verbal and nonverbal social behavior.

These findings are consistent with a large body of anecdotal and observational data attesting to the beneficial impact of animals on psychological, social, and physiological indicators with diverse populations. One cautionary note should be mentioned. Our study involved only nineteen (19) patients and spanned 10 weeks. Future investigators should attempt to use larger numbers of subjects and larger periods of follow-up.

References

Brickel, C.M. (1979). The therapeutic roles of cat mascots with a hospital-based geriatric population. The Gerontologist, vol. 19, no. 4, pp. 368-372.

Corson, S.A. and Corson, E.O. (1982). Pet animals as socializing catalysts in geriatrics: an experiment in nonverbal communication therapy, in Society, Stress, and Disease, vol. 5, ed. Lennart, Levi (Oxford: Oxford University Press).

Corson, S.A., Corson, E.O., and Gwynne, P.H. (1975). Pet-facilitated psychotherapy, in Pet Animals and Society, ed. R.S. Anderson (London: Bailliere Tindall).

Daniel, S.A., Burke, J., Burke, J., Comprecht, J. and McLaren, T. (1988). The effects of an animal-assisted therapy visitation on emotionally disturbed youth. Presentation of abstract to Delta Society Conference, Orlando, Florida.

Daniel, S.A., Burke, J., Rutel, Y., and Burke, J. (1987). Physiological effects of a pet visitation program: comparisons of different care levels. Presentation of abstract to Delta Society Conference, Vancouver, British Columbia.

Friedmann, E., Katcher, A.H., Lynch, J.J., and Thomas, S.A. (1980). Animal companions and one-year survival of patients after discharge from a coronary care unit. Public Health Reports. pp. 95, 307-312.

Katcher, A.H. (1982) Are companion animals good for your health? A review of the evidence. Aging, pp. 2-8.

McCulloch, M. (1981) The pet as prothesis: defining criteria for the adjunctive use of animals in treatment of medically ill, depressed outpatients, in Interrelations Between People and Pets, ed. Bruce Fogle (Springfield, Il: Charles C. Thomas).

Mugford, R.A. and M'Comisky, J.G. (1975). Some recent work in the psychotherapeutic value of caged birds with old people, in Pets, Animals and Society, ed. Robert S. Anderson (London: Bailliere Tindall), pp. 54-65.

Muschel, I.J. (1984). Pet therapy with terminal cancer patients. Social Casework, vol. 65, no. 8, pp. 451-458.

Robb, S. Boyd, M., and Pritash, C.L. (1980). A wine bottle, plant, and puppy: catalysts for social behavior. Journal of Gerontological Nursing, 6, pp. 721-728.

Siegel, J.M. (1990) Stressful life events and use of physician services among the elderly: the moderating role of pet ownership. Journal of Personality and Social Psychology. Vol. 58, No. 6, pp.1081-1086.

Appendix A

BUILDING MENTAL HEALTH

Self-Esteem

Having a good opinion about ourselves means having high self-esteem. High self-esteem is one of the most important qualities of mental health. High self-esteem can make you feel capable, productive, and self-confident. With high self-esteem, you can act more like the person you want to be. You can enjoy other people more fully and can give to other people. High-esteem can give you the confidence to try new activities. It can help you feel happier with yourself.

Low self-esteem can hurt us in many ways. With low self-esteem, we lose confidence in ourselves. We may feel doomed to fail again. This belief may actually cause us to make little or no effort in reaching the goals that are important to us. Low self-esteem causes us to see other people as more successful, smart, or more important than ourselves. As a result, we may start to believe that good things only happen to other people and couldn't happen to us. We start to feel like victims. This further reduces our self-esteem.

HOW TO INCREASE SELF-ESTEEM

Identify your strengths

One of the first steps in increasing self-esteem is determining those areas in which you have a special ability, Some people are good in sports, whereas others like to cook. Some enjoy taking care of pets but others have mechanical ability; they can make repairs on cars, for example. Other people can play a musical instrument or sew.

It doesn't matter what skill you excel in. Recognize and accept your strength no matter what it is. Be proud of it. It's your special area and nobody can take that away from you.

Accept Your Limitations

Nobody is perfect and nobody is good at everything. Some people have a warm personality but can't balance a checkbook. Others are good at arithmetic and have common sense but lack confidence. Each of us has some area of strength but we all have limitations. Don't expect to be the best. Very few people are recognized for being the best in their field of interest. If you're not as smart as your friend or relative, try to accept this limitation. You may not become famous or rich, but you still can make a valuable contribution to life. Take pride in what you do and don't reach for the unattainable. Doing this will only make you feel like a failure.

Set Realistic Goals

Most successful people have had to work hard to accomplish their goals. It's easy to say, "I'm going to become the greatest actor in the world." Accomplishing that goal may be impossible if you have never taken an acting lesson or never went on an audition, or haven't hired an agent, or never acted in a play, show or movie. The first step in achieving a goal is determining if it is realistic. If you never went to an acting class, chances are you'll never become a great actor. If you never played baseball, it is unlikely, probably impossible, that you'll play in the World Series one day. If you want to work and haven't worked for 5 years, it is more realistic to attend a workshop or business practices program than to look for a job by reading the "want" ads. If you want to be a typist and can type only 10 words/minute, it is better to practice your typing skills than to go on job interviews. If you want to live in an apartment, it is better to learn how to cook, and do the laundry than to immediately sign up for a supportive apartment program. People succeed by taking small steps towards the goals that are important to them. Take one step at a time and don't move to the next one until you've mastered the first. Reaching too high and too fast will produce failure.

Stop Saying, "I Can't"

It is natural for a person to get discouraged if after hard work and effort he still hasn't reached his goal. You may have set your sights on becoming a secretary, a nurse, a mechanic, or perhaps a doctor, lawyer or an engineer like your older/younger brother or sister. Instead, you've spent many years in workshops, vocational training centers, or in jobs lasting only a few weeks or months. Progress was often interrupted by the onset of voices or paranoid feelings which caused a hospitalization. It is natural to feel discouraged and disappointed in yourself.

Don't give up hoping. Stop saying, "I can't!" Start saying, "I can't become a (e.g., doctor, nurse) but I can become a _____." " I can't make $50,000/year, have a house in the suburbs, get married and have 3 children, but I can earn spending money every week, live in a clean and safe residence and have friends."

Taking a "can do" attitude will give you the encouragement and also help you strive for goals that are respectable but also realistic.

Building Pride

Take pride in your achievements both large and small. Remember the statement, "I did a good job." These are important words. As an adult being proud of yourself and your accomplishments should be more important than the words of encouragement you receive from family, friends and, yes, even your therapist. We all want others to recognize our accomplishments. It's natural to feel this way and, yes, it feels good when people who we respect compliment us or tell us we're doing a good job.

Just as accomplishments and praise feel good, we have to start accepting and believing the words or praise we give ourselves. "There, I did a good job, it means I'm a good person. I try hard, I'm a person of value, I'm worthy and decent, and I'm lovable." Try to tell yourself this and <u>believe it</u>. It's true. Focus on positive and learn to ignore the negative statements we sometimes tell ourselves. Focusing on the negative leaves little room for positive and in order to feel good about ourselves, to have self-esteem, we need more positive than negatives. Remember, your accomplishments are your own. Be proud of them and yourself.

Appendix B

Coping With Mental Illness

Maintaining Mental Health

Increasing self-esteem is an important ingredient in becoming mentally healthy. However, it's not the only ingredient. What is mental health? Just like mental illness, mental health has many parts. Few people are mentally healthy in all areas, but it is important to know what those areas are. Here's a list of ingredients that mental health workers and patients with mental problems have identified.

1. Well-Being: All people have problems and problems are a normal part of living. Nevertheless, people who are mentally healthy report feeling O.K. about their lives and themselves. In spite of their day-to-day problems, they feel at peace with themselves and seem to possess an inner calmness.

2. Self-Confidence: As we learned, an important ingredient in self-esteem is self-confidence. People who are mentally healthy believe in themselves and their capabilities. They may go through periods of insecurity and self-doubt, but usually they are confident in their ability to solve life's problems.

3. Fulfilling Responsibilities: Being reliable, self-disciplined and motivated are all necessary in completing life's tasks, whether the task is waking up on time, meeting a friend at a specific place, answering questions at a day program, packaging at work center, showering on a daily basis or earning tokens to move up in the levels at the residences, people who are mentally healthy must demonstrate their ability to carry out responsibilities. Fulfilling our responsibilities tells others that we are valuable, hard-working and trustworthy. It tells people they can count on us.

4. Remaining Optimistic: It's hard to feel optimistic about the future if the present and past have been a struggle. Yet, mental health means remaining positive about the future. It means having hope and believing in a better future. We all hope but it is <u>up to us</u> to keep believing that our lives will improve. Some people find hope through organized religion. Others have a spiritual sense even if they never attended church, synagogue or a religious service. To believe in a better future, we must believe in ourselves. Even if a person can't accomplish his most cherished goals, he can still try to improve in other areas. Where there is effort, there is hope.

Remember, thinking in negative terms or being pessimistic will only make the future harder than it has to be. Experts tell us negative thinking (expecting a future of gloom and doom) will make it hard to succeed. Think positively. Even if you believe your life is controlled by fate or luck, think about ways in which you can control events in your life. Positive thinking is not magic. It means telling yourself that things will get better, that bad luck can turn to good luck, that life has some beautiful aspects, that in each day there can be some joy, some laughter, some calm, no matter how brief.

When your head is filled with negative thoughts and bleak feelings, talk to yourself. Tell yourself, "This too shall pass," "Tomorrow is another day," "If I had a poor voucher this morning, I'll earn a better one this afternoon." Being mentally fit means actively trying to make things better.

5. Freedom for Anxiety: This is a tall order. As we learned in stress curriculum, it is impossible to be human and to be completely free of stress. Too much stress is bad. It can cripple us: stop us from carrying out our responsibilities, stop us from taking care of personal hygiene, stop us from making friends. Too little stress can make us lazy, without goals, stagnant. Some stress or some anxiety will motivate us to make an effort to work toward our goals. If you feel anxious, try to distract yourself with a hobby or a favorite interest, read a book, go for a walk, watch T.V., talk to a friend, do some sewing, clean your room, think about pleasant thoughts. Remember, if you focus away from what is distressing you (making you anxious), even for a short period, you will be able to break that cycle of anxiety.

6. Concentration: People who are mentally fit report they can concentrate on whatever activity they are currently involved in. Concentration is extremely difficult when voices intrude, memories from childhood creep into our thoughts or past mistakes are repeated in our thoughts like a broken record. It is extremely hard to finish a puzzle, listen to a conversation or pay attention when so many thoughts or feelings (like guilt) keep flooding our minds. Medication can help us concentrate and so can we. Reach out to someone whom you trust when the inner distractions become very strong. Starting a conversation with a friend, you may be able to lessen the impact of the critical voice(s) in your head. Playing a game of pool or visiting a museum or taking in a movie can take your mind off those troubled feelings. Remember, one of the keys to improving concentration is to take the focus away from yourself and place it on something or someone else.

7. Keeping Problems in Perspective: All people have problems. Problems are a natural part of life. Problems can only threaten one's mental stability if they pile up one on top of the next. Nobody can solve several problems at the same time. Most of us can only pay attention to one problem at a time. By focusing on several problems at once you will not be able to solve any of them. Pretty soon, you'll feel overwhelmed, anxious, angry or depressed.

For example, you may wake up late one morning and discover the bathroom is occupied. You have 25 minutes to get to program and you haven't dressed yet. You remember that you're supposed to speak to a friend to arrange a trip to K-Mart after the program. What should you do first? Don't waste your time getting upset about the bathroom situation. Either get dressed or go to another bathroom. After you've dressed, go back to the bathroom to see if it is available. Don't speak to your friend until you have free time at the Day Hospital.

Remember, don't try to solve all your problems at once and try to stay calm. Getting angry or panicky will not solve your problem of lateness. Solve one problem at a time and go to the next.

8. Be in Control (Staying in Control): People who are in good mental and physical health are able to exert control over their lives. Whether deciding what to eat, how to spend their money, when to take a shower, or what kind of clothing to wear, people feel best when they can make choices. Freedom of choice is our right and should be respected as long as we exercise sound judgment.

What do you control in your life? What choices do you make on a daily basis? Why aren't you in control of what you want to do? What's getting in the way? Remember, if we want to be more in control over the way we live our lives, we must be aware of the consequences of our choices. If I live in CRSP, will I attend regularly or will I skip days at my new program? If I decide to live in Family Care, will I agree to do chores at my new home?

Making choices and making changes in our lives means thinking ahead to all the new responsibilities that a new choice entails. It's easy to say, "I want to leave the CRTS Day

Hospital. I want to live at Beacon House I, in Family Care, back in Brooklyn with my family, or in a supportive/supervised apartment." It's more difficult to list all the new responsibilities and adjustments that will automatically occur and then come up with a plan of action for solving these problems.

Effect of Pets on the Well-Being of the Elderly

Cathleen M. Connell, M.S.
and
Daniel J. Lago, Ph.D.

That social contacts and involvement are important to the well-being holds much intuitive appeal. For most of us, life without the rich and varied experiences provided by and shared with family, friends, and neighbors is an impoverished one indeed. Recently, studies on the factors that contribute to mortality lend empirical evidence to the vital roles of social participation in the lives of the elderly. House, Robbins, and Metzner (1982) discovered that individuals reporting higher level of social relationships and activities (adjusting for age and a variety of risk factors) were significantly less likely to die in the follow up period of the community health study involving a cohort of 2754 adult men and women. These results were invariant across occupational, age and health social status groups. Thus, the benefits of social ties have both intuitive and empirical support. For some elderly individuals, however, a constricting life space results in reduced availability and opportunity for social participation. Such change may alter the impact of social participation on the morale and life satisfaction of the elderly. It is suggested, therefore, that the notion of social participation be expanded to include other forms of involvement.

The present investigation was undertaken to assess the impact of a previously overlooked source -- the contribution of companion animals. This relationship may be particularly significant for elderly individuals who live alone or have experienced a serious social or personal disruption in later life. For these individuals, the companionship, support, caring and intimacy thought to accompany the marital relationship may not be readily available. Thus, a personal coping resources, such as a relationship with companion animals, may serve as influential mediating factors in determining what contributes to the life satisfaction of the elderly.

A secondary analysis of the data from the Companion Animal Project being conducted at The Pennsylvania State University (Lago, Knight & Connell, in press) was performed to simultaneously address two hypotheses: (1) Among respondents who currently own a pet, a favorable attitude toward that pet significantly contributed to reported happiness, (2) Among respondents who are widowed, divorced never married, are separated, a favorable attitude toward a pet contributes more to the variance in perceived happiness than among married respondents.

THE DATA AND SAMPLE

The data are drawn from the Companion Animal Project, a four-year cross-sequential research project being conducted to determine whether owning pets effects the health and well-being of older persons. Structured personal interviews, including sections on animal ownership, attitudes towards pets in general and toward a particular pet, social activities, social satisfaction, physical health, activities of daily living, emotional health, and demographic information were administered to 184 elderly respondents residing in Centre County, Pennsylvania. The sample was obtained through both a group and door-to-door recruitment approach. The present investi-

gation is based on the data collected from pet owners in the second time interval of the study. Interviews averaged approximately 1 1/2 hours in length. The 184 respondents (80 of whom are pet owners) range in age from 54 to 90, with a mean age of 70. Forty males and 144 females were interviewed, 182 of whom were white, two black. The Approximate average income for the sample was $9,000 a year. Twenty-three of the respondents were never married (12.5%), 76 are married (41.3%), 75 are widowed (40.7%), seven are divorced (3.8%) and three are separated (1.6%).

DEPENDENT AND INDEPENDENT VARIABLES

Self-reported Happiness, the Dependent Variable, is measured by an adapted form of the MUNSH (memorial University of Newfoundland Scale of Happiness)(Kozma and Jones, 1980). Eight of the items from the Pet Attitude Scale (PAS) devised by Templer (1981) were chosen to measure a Favorable Attitude Toward A Pet.

Four additional independent variables were included in the present investigation. Income was assessed by a single indicator: self-reported annual income. Measures of Self-Rated Physical Health, Social Behavior, and Social Satisfaction were constructed from unweighted composite sums of several items included in the interview schedule. Functional Status was assessed by use of eleven items derived from the Activities of Daily Living section of the Older Americans Resources Survey (OARS) (Duke University, 1978).

In order to determine the relative contribution of a favorable attitude toward a pet, the previously described independent variables which have been consistently found to be predictors of well-being among the elderly in previous social gerontological research were included in the analysis. Income, self-rated physical health, activities of daily living, social behavior, and social satisfaction, in addition to a favorable attitude toward a pet, were entered into a series of step-wise multiple regression procedures to determine their ability to predict perceived happiness in this sample of rural elderly.

In order to determine whether a favorable attitude toward a pet will contribute more to perceived happiness among unmarried respondents than married respondents, comparisons of its contribution were made between the two subgroups.

RESULTS

A favorable attitude toward a pet does contribute to the predictive power of a model used to explain variance in the happiness score of a sample of rural elderly who own pets (see Table 1). The increment of small, though in the direction predicted (beat value = 3.15). It explains relatively more of the variance in MUNSH scores than does social satisfaction or activities of daily living (both of which were dropped from the model because they did not reach levels of statistically significant prediction). Self-rated physical health was found to be the most important predictor of perceived happiness, followed by social behavior and income.

Table 1

Stepwise Prediction of MUNSH Scores for Pet Owners (n=80)

Predictor	Beta Value a	Significance
Self-Rated Physical Health	16.18	.0058
Social Behavior	13.49	.0610
Income	13.35	.0321
Favorable Attitudes Toward Pets	3.15	.0357

$R^2 = 0.2369$ $F = 5.82$ $2 < .0004$ a = Standardized partial regression coefficient

A favorable attitude toward a pet also emerges as a significant predictor in the models for both the pet-owning married and the unmarried respondents (see Table 2). For the unmarried respondent, a favorable attitude toward a pet is positively associated with perceived happiness (beta value = 8.37). However, for married respondents, the beta value associated with a favorable attitude toward a pet is negative (-1.25). This indicates that those individuals who report more favorable attitudes towards pets tend to score lower on the indicator of perceived happiness.

Table 2

Stepwise Prediction of MUNSH Scores by Marital Status for Pet Owners Only (n=80)

Predictor	Beta Value a	Significance
Married Respondents (n=33)		
Social Behavior	22.0	.0252
Self-Rated Physical Health	11.21	.0037
Favorable Attitude Toward Pets	-1.25	.0438
$R^2 = 0.3913$	F = 6.21	2 < .002

For the unmarried pet-owning group, self-rated physical health is the single best predictor of happiness scores in this sample, followed closely by income. This variable explained twice as much of the total variance as did self-rated physical health, which was entered second. Most of the total variance in happiness scores can be explained for the married group than for the unmarried group (39.13% compared to 14.44%)

Widowed, Divorced, Never Married and Separated Respondents (n=47)		
Self-Rated Physical Health	15.92	.1149
Income	16.28	.1125
Favorable Attitude Toward Pets	8.37	.2628

$R^2 = 0.1444$ F = 2.42 $2 < .079$ a = Standardized partial regression coefficient

DISCUSSION

In light of the changes that take place in the social world of some elderly individuals and the alleged benefits of pet ownership, a positive disposition toward a pet was thought to be a personal form of coping and social involvement previously overlooked in gerontological research. The present investigation was undertaken as an empirical test of this perspective.

In general, a favorable attitude toward a pet was found to be predictive of perceived happiness among pet owners. In fact, in the regression models used for pet owners, a favorable attitude toward a pet contributed more to perceived happiness than several traditionally well-documented factors (i.e., social satisfaction and activities of daily living).

A favorable attitude toward a pet was positively associated with perceived happiness for the unmarried respondents but was negatively associated with perceived happiness for the married individuals in the sample. Additionally, a favorable attitude toward a pet explained a higher percentage of the variance in perceived happiness among unmarried respondents than among married respondents.

These findings support the contention that animal companionship may play a more important social role in the lives of those elderly who live alone. More importantly, they suggest that the effects of a close relationship with a pet may not be uniformly beneficial for all elderly individuals. A pet might impede marital closeness and become a source of conflict in some households. This finding is further supported by data indicating that animal placements which were unsuccessful are twice as likely in multiple-person households (PACT'S initial evaluation, in press). Often the pet ownership and caregiving styles of each spouse differ. Partners may unwillingly be forced to provide care for an infirmed spouse's pet. With the introduction of a pet in a formerly closed dyadic relationship, there may be a risk that one partner will feel excluded or burdened. Additionally, the married respondents reported higher levels of social activity and tended to be younger than the unmarried respondents. These factors contribute to the more central role of pets in the lives of the unmarried individuals.

The need to determine more specifically what benefits individuals derive from their relationships is highlighted. Constancy, open interaction, and physical contact may each play a role in relationships with people and animals. More detailed information is also required to assess the quality of a relationship with a pet and the circumstances under which it becomes rewarding and beneficial or a source of conflict.

In the present study, the findings discussed were based upon a relatively small sample of rural Pennsylvania elderly. However, tentative suggestions for interventions involving pets can be offered as a result of the findings of this study. All members of households involved in a pet-placement effort need to be consulted at the intervention's conception. Responsibility for care and ownership styles should be considered as well as the potentially disruptive effect of pets on relationships in the household. When the pet is being placed for the benefit of one person, the concerns and reactions of each should be integrated into a plan to maintain the placement. Also, this study confirms the notion that the unmarried, widowed elderly who live alone and accept or seek a pet may benefit greatly from their relationship with a companion animal.

Finally, additional support is given to the growing recognition that animals are a significant addition to the social world of many elderly individuals. The preliminary findings of this investigation will be clarified and strengthened if the role of pets is included in future research on gerontological social relationships.

References

Duke University Center for Study of Aging and Human Development. Multidimensional functional assessment. The OARS methodology (2nd ed.). Durham, NC: Duke University Press, 1978.

House, J. Robbins, C, & Metzner, H. The association of social relationships and activities with mortality. Prospective evidence from the Tecumseh Community Health Study American Journal of Epidemiology, 1982, 116, 123-140.

Kozma, A. & Stones, M.J., The measurement of happiness: Development of the Memorial University of Newfoundland Scale of Happiness (MUNSH) Journal of Gerontology, 1980, 35, 906-912.

Lago, D., Knight, B. & Connell, C. Rural elderly relationships with companion animals. Feasibility study of a Pet Placement Program. In A.H. Katcher & A.M. Beck (eds) Proceedings of the International Conference on the Human/Companion Animal Bond. Philadelphia, PA: University of Pennsylvania Press.

Templer, D. The construction of a pet attitude scale. Psychological Record, 1981, 31, 343-348.

The Latham Letter, Spring, 1983, Vol. IV, No. 2, pp. 14-16.

Animal Assisted Therapy in the Psychiatric Setting

Christine Shaheen, AAT

BACKGROUND OF THE STUDY

The San Francisco SPCA's Animal Assisted Therapy (AAT) Program has grown from a recreational activity (formally established in 1981) to a valuable therapeutic intervention which is now widely in demand (currently, our waiting list numbers over 30 facilities). The Program serves approximately 24,000 clients per year in over 150 San Francisco sites.

A survey was developed by the author in September of 1988 to collect data on the following aspects of the SPCA's AAT program: volunteers, animals, visit format and effects. The in-depth survey sought to identify some of the issues inherent in the "therapeutic value" of AAT from the standpoint of professional staff working with recipient clients. These professionals are in the strongest position to evaluate what the benefits are.

A pilot survey was sent to the staff responsible in three representative facilities: an acute care pediatric unit, an adult locked psychiatric ward as well as a board and care retirement home. The main survey was sent to 76 of our most frequently visited sites (visits range from twice a month to once a year) in December of 1988. Forty-two facilities responded, a return of 55%. Respondents to the survey include staff from the following: skilled nursing, adult day health, hospital acute care, senior citizens/nutrition sites, rehabilitation/spinal injury units, a special school and a day program for developmentally disabled individuals.

The following is a summary of responses from 10 psychiatric facilities which represents 24% of all survey returns (psychiatric and skilled nursing facilities comprise the two largest group of clients served by the SF/SPCA Program). The respondents comprised three groups: hospital unlocked units, hospital psychiatric unlocked units and day treatment psychiatric units. While responses were similar from all three groups, a few significant differences were indicated.

Although some of the earliest records on the involvement of animals in therapeutic programming originate in the psychiatric setting (from York Retreat of 1792 to the pioneering work of Drs. Samuel and Elizabeth Corson in the 1970s), there is relatively little data concerning the specific benefits of AAT from persons suffering from chronic or acute illness.

PROFILE OF CLIENTS/STAFF/FACILITY

Staff respondents reported having animals at home: 70% of them do not. It may be assumed that the age of clients ranges from five to 75 plus years (some respondents elected not to answer the client age question), since this reflects the ages of the overall psychiatric population served by the Program.

THE VISITS

All locked units reported on received two visits per month but desired four visits. The other two groups expressed satisfaction with the visiting frequency (once or twice a month). One hundred percent of the respondents reported that visits were conducted in group activity rooms. The number of staff participating in the AAT groups ranged from one to four, depending on the group size. Other participants included volunteers, visitors and student interns.

THE ANIMALS

Currently our AAT Program visits include several species of animals housed in the SF/SPCA Education Resource Center. The personal animals of some volunteers and staff are also involved (mostly dogs and an occasional cat or bird). Facility staff were asked which animals were the most and the least popular at their respective sites. In all three groups, dogs were the favorite animal reflecting 80% of the responses. Guinea pigs were second favorite with a 40% response, and owl and rooster both rated 20%. Rats, iguanas, cats, birds and snakes were also noted as popular with clients. Some respondents listed favorite animals by name (i.e., "Mavis the dog").

Even more variety was reflected in the least favorite category. Of nine respondents 66% reported rats as least favorite, followed by snakes at 44% and lizards at 33%. The rooster, cats and hamsters were also listed as least favorite.

ATMOSPHERE

Nine respondents reported that the atmosphere at their facilities changed when the AAT visits took place. The most frequently reported changes were that the atmosphere became "more lively" (44%) with "more excitement" (44%), "more interest and attendance" (44%) and "greater interaction" (33%). Other respondents remarked that the unit became "less institutional and more family-like," that "withdrawn patients express themselves" and there is a "show of affection," and the atmosphere becomes "warm, intimate and non-threatening."

MAIN BENEFITS

"If the clients visit with each other during pet therapy we see this as a positive reaction, since they often just sit and stare without any socialization. Conversations with each other are often centered around animals and reminiscence."

Among all three groups of psychiatric clients, the main benefits of AAT reported were socialization, reminiscence and touch (90% each). These were followed by contact with the natural world (60%) and reality orientation (50%). Also reported were relaxation (40%), distraction (30%) and lower blood pressure (10%).

It is significant to note that clients on locked psychiatric units have the least amount of contact with the outside or natural world. Those in each of the other treatment settings have the opportunities to leave the setting or live in the community. Although each group considers contact with the natural world important this received a 100% rating as a main benefit only by the staff on the locked units. And while socialization was rated as a main benefit of both the locked and unlocked groups, it was only rated as a main benefit by 80% of those working in the day-treatment setting.

WHO BENEFITS MOST?

Of nine responses to this question, 66% considered the depressed client as deriving the most from AAT. Thirty-three percent saw the isolated or withdrawn client as benefiting the most, and 22% thought demented and nonverbal clients derived the greatest benefit.

ARE THERE CLIENTS WHO DISLIKE THE ANIMAL VISITS?

There were four responses to this question: on locked units, paranoid patients sometimes feel threatened by the animals; on unlocked units, psychotic or hallucinating patients may be fearful; in day-treatment facilities, animals may be disliked because of fear of allergies. Sixty percent of the respondents stated that no patients disliked the animal visits.

HOW DOES AAT DIFFER FROM OTHER FORMS OF THERAPY?

This question drew mixed responses. The largest percentage of respondents (44%) indicated that AAT differed because it was safe, less threatening and less performance oriented than other therapeutic activities. Thirty-three percent of respondents said it differed because it was a form of stimulation or because it was conducted by someone other than facility staff.

LONG-TERM EFFECTS OF AAT

Overall responses indicate the long-term effects of AAT include anticipation of the visits, reminiscence, greater attendance at activities and caring for others. However, one respondent from both the locked and unlocked categories indicated there were no long-term effects because socialization decreased as soon as the visits were over, and patients became withdrawn.

DOES AAT STAFF MEET TREATMENT GOALS FOR PSYCHIATRIC PATIENTS?

"The visits improve motivation for attending groups and working on treatment for (the clients) psychiatric problems." Ninety percent of the staff responded positively to this question and cited the following reasons: increased socialization (77%) and increased attendance for therapeutic activities (55%). Other comments on how AAT helps meet treatment goals include improved affect (mood), increased reality orientation, development of social skills and motivation to work on further psychiatric treatment.

GENERAL COMMENTS: VOLUNTEERS, ANIMALS, VISIT FORMAT

"In a very short time, I have seen and learned more about our clients in this one group than in any other interactive group because of the animals."

A wide variety of comments and suggestions by respondents from all three groups were submitted with the surveys. Following is a brief summary.

ABOUT THE VOLUNTEERS

By sharing information about the animals, "they create a learning experience for the clients and me." By being sensitive to the needs of clients, the volunteers can help "engage and motivate (clients) to participate." It helps when the volunteers are on time because "patients get disinterested if they have to wait." It helps to work "one-on-one" with the more withdrawn clients.

ABOUT THE ANIMALS

The clients "often have similar stories (to the animals') about being homeless." Bringing a variety of animals "helps keep the clients interested" and provides a "learning experience." Bringing two animal visitors of different species works especially well. Clients enjoy most "playing games with and feeding the animals." "Repeat visits by the same animal" was recommended.

THE VISIT FORMAT

It is helpful for the AAT volunteers to "structure the (visit) and state clear guidelines ... regarding how to interact with the animals." "Structure is important but there should be room for spontaneity." Also helpful is giving the clients "lots of latitude in how much contact to have" with the animals.

SUMMARY AND RECOMMENDATION

The psychiatric staff who responded to the survey are professionals whose client groups have on-going AAT visits. On the average, they have worked in their facilities for a minimum of one and one-half years. Thus we may conclude that the Program is well established and highly valued in their respective facilities. Interest in supporting and improving the Program may have motivated them to respond to the survey in such large numbers.

A logical follow-up to the staff survey would be an in-depth survey of the AAT clients themselves. Although respondent professionals are commenting on behalf of the clients they work with, staff members' own preferences and biases may be reflected in these data. Other aspects of this survey will be reported on in future issues of The Latham Letter.

For further information or copies of the AAT survey, contact the author at: SF/SPCA, 2500 16th Street, San Francisco, CA 94103.

Chris Shaheen, MA is now a private consultant, lecturer and author on AAT. This study was conducted while she was the AAT Program Specialist at the SF/SPCA.

For copies of the survey, contact the SF/SPCA Education Department at 2500 16th Street, San Francisco, CA 94103.

The Latham Letter, Vol. X, No. 2, Spring, 1989 pp. 1, 13-14.

Stressful Life Events and Use of Physician Services Among the Elderly: The Modifying Role of Pet Ownership

Judith M. Siegel, Ph.D.

The physician utilization behavior of 938 Medicare enrollees in a health maintenance organization was prospectively followed for 1 year. With demographic characteristics and health status at baseline controlled for, respondents who owned pets reported fewer doctor contacts over the 1-year period than respondents who did not own pets. Furthermore, pets seemed to help their owners in times of stress. The accumulation of prebaseline stressful life events was associated with increased doctor contacts during the study year for respondents without pets. This relationship did not emerge for pet owners. Owners of dogs, in particular, were buffered from the impact of stressful life events on physician utilization. Additional analyses showed that dog owners in comparison to owners of other pets spent more time with their pets and felt that their pets were more important to them. Thus, dogs more than other pets provided their owners with companionship and an object of attachment.

Current and projected demand by the elderly for health care services has prompted the study of their physician utilization behavior. These studies and others show that factors in addition to physical health status influence decisions to use medical services. Psychological stress, for example, has been positively associated with the frequency of primary care physician visits among the general population (Barsky, Wyshak, & Klerman, 1986; Regier, Goldberg, & Taube, 1979; Shuval, 1970; Tessler, Mechanic, & Dimond, 1976) and among the elderly (Waxman, Carner, & Blum, 1982). Stressful life events also contribute to higher utilization rates (Rahe & Arthur, 1978) because stressful life events are intertwined with psychological distress and because persons undergoing stress pay greater attention to bodily symptoms as well as find them more disturbing (Mechanic, 1972). One of the most distressing life events, death of a spouse, occurs with greater frequency in older populations. Major events, such as spousal loss, are frequently identified as precipitating factors in loneliness (Perlman & Peplau, 1984), another potential determinant of physician utilization. In light of these notions, it is reasonable to hypothesize that circumstances that promote well-being or alleviate distress or both could reduce the need for physician contact. One such circumstance is pet ownership, as pets have been reported to provide companionship, an aid to health and relaxation, protection, and nonjudgmental acceptance and love (Soares, 1985).

A rich anecdotal lore exists in support of pets as companions to the elderly, although methodically strong empirical studies are few. Observational studies suggest that introducing pets into the lives of terminal cancer patients (Muschel, 1984) or the lives of patients in a geriatric ward (Brickel, 1986) brings about significant positive social and psychological consequences. Bird placement among British pensioners led to positive psychological effects in comparison with pensioners who received a plant (Mugford & M'Comisky, 1975). However, at least one evaluation of a companion animal program failed to show positive gains for those who acquired pets relative to a comparison group (Lago, Connell, & Knight, 1983); among pet owners, though affection for pets was positively related to morale.

With regard to naturally occurring pet ownership, one study (Robb & Stegman, 1983) found no physical benefit and three studies found no psychological benefit (Lawton, Moss & Moles, 1984; Ory & Goldberg, 1983; Robb & Stegman, 1983) of pet ownership among the elderly. This is in contrast to dramatic findings that identified pet ownership as a strong social predictor of 1-year survival in a group of postcoronary patients (Friedmann, Katcher, Lynch & Thomas, 1980). Also supportive of the value of pets are data from a national probability sample of respondents 65 years or older that showed pet attachment was inversely related to depression as measured by a symptom scale (Garrity, Stallones, Marx & Johnson, 1989). Furthermore, Garrity, et al., found that pet attachment was associated with better physical health, as assessed by retrospective reports of recent illness experiences among respondents with low levels of human support but not among those with adequate human support. These data suggest that pet ownership or attachment or both might play a beneficial role in times of stress. Similarly, data on human social relationships can buffer the impact of a variety of stresses and strains such that individuals experiencing high stress and high social support will not evidence compromised physical or psychological health (Broadhead, et al., 1983).

This study prospectively examined the direct and indirect effects of pet ownership on utilization behavior of the elderly. Specifically, I hypothesized that with demographic and health characteristics controlled for, pet owners would report fewer doctor contacts than non-owners. I did not anticipate a relationship between pet ownership and doctor contacts among respondents in low-stress circumstances. In this study, stress was operationalized in two ways: depressive symptomatology and accumulation of stressful life events.

METHOD

Sample

The data for the present investigation were collected as part of a 1-year panel study concerned with health behavior of the elderly. All respondents were enrolled through Medicare in a federally qualified network model health maintenance organization (HMO) located in southern California. At study outset, approximately 2,900 members of the HMO were enrolled through Medicare. After eliminating one member of spouse pairs from the sampling frame, study solicitations and consent forms were mailed to approximately 2,300 potential participants. Signed consent forms were returned by 1,145 enrollees. Data collected earlier by the HMO permitted a comparison of enrollees who returned the consent forms and those who did not. The two groups were comparable in age, gender compositions, marital status, and self reports of depressed mood. However, the respondent group differed from the nonrespondents in terms of greater representation of non-Hispanic Whites, high school graduates or above, and good or excellent health status.

Interviews were conducted by telephone unless poor hearing or other impairment on the part of the respondent interfered. In these instances, interviews were conducted face to face. In total, 1,034 respondents 65 years of age or older were interviewed at baseline (58 interviews were conducted face to face). Among the 1,145 potential respondents, attrition was due primarily to death, severe illness, relocation, refusal by another family member, or reported age of less than 65 years. The baseline questionnaire assessed health status, health beliefs, psychological distress, social network and support, pet ownership, and demographic characteristics.

Every 2 months for the 12-month period following the baseline interview, respondents were re-interviewed concerning doctor contacts that had occurred since the prior interview. The measures of psychological distress were repeated at Wave 4 (6 months) and at Wave 7 (12 months). For pet owners, information on the nature of their relationship with their pets was collected at Wave 2.

MEASURES

At baseline, respondents reported whether they had "any chronic, that is, recurring or continuing health problems," provided data on their demographic characteristics, and answered questions concerning the extensiveness of their social network. Demographic characteristics assessed were gender, age, racial-ethnic group, income, education, marital status, employment status, and current living arrangement. Included among the question on living arrangement was one asking whether there were any pets in the household and, if so, what type(s). Social network involvement was assessed by the 10-item Lubben Social Network Scale (LSNS; Lubben, 1988), developed for gerontological research. This scale has three components: family networks (items are "number seen monthly," "frequency of social contact," "number respondent feels `close to,'") friendship networks ("number seen monthly," "frequency of social contact," "number feels `close to'") and interdependent social supports ("has a confidant," "is a confidant," "relies upon and helps others," "living arrangement"). A total LSNS score is achieved by summing the 10 items, each of which ranges in values from 0 (least connected) to 5 (most connected). Criterion-based validity and internal consistency are adequate (Lubben, 1988).

At baseline, at 6 months, and at the final interview (12 months), depressed mood was measured by the Center for Epidemiologic Studies Depression Scale (CES-D; Radloff, 1977). The CES-D uses 4-point scales (scored 0 to 3) to assess the frequency with which each of the 20 symptoms was experienced during the previous week. Several epidemiologic studies (Comstock & Helsing, 1976; Frerichs, Aneshensel, & Clark, 1981; Husaini, Neff, Harrington, Hughes & Stone, 1980) found the scale sensitive to differences in level of depressed mood. The CES-D has adequate validity (content, criterion-based, and construct) and reliability (test-retest and internal consistency; Radloff, 1977; Weissman, Sholomskas, Pottenger,

Prusoff & Locke, 1977) and has been used previously with aged populations (Berkman, et al., 1986; Garrity, et al., 1989; Murrell, Himmelfarb & Wright, 1983).

The measure of life events, administered at baseline 6 months, and 12 months, was a combination of checklists developed for gerontological populations (Amster & Krauss, 1974; Kahana & Kahana, 1983; Lubben, 1984). Respondents indicated whether any of the following 10 events occurred to them in the previous 6 months: separation or divorce, death of a close family member, major illness of spouse, job retirement, death of a close friend, move, being a victim of crime, being denied a driver's license, and money problems.

Use of physician services was assessed every 2 months (at Waves 2 through 7) and aggregated across the waves. The respondents were asked how many times they had contacted the doctor since the last interview. Respondents reporting at least one contact were asked how many times they specifically requested to see the doctor and how many times the doctor specifically requested to see them.

Respondents identifying themselves at Wave 1 as pet owners were asked a series of questions at Wave 2 about their pets. Respondents having more than one pet were asked, "Which one of your pets is your favorite one? That is, the one to whom you are the closest or give the most attention?" All subsequent questions referred to the respondents' favorite pet (or only pet).

Four aspects of the human/pet relationship were assessed: responsibility, time with pet, affective/attachment to pet, and benefit minus cost difference. Two questions assessed responsibility: Whose decision was it to get the pet? And who is most responsible for the care and feeding, including trips to the veterinarian? Responses were categorized as respondent alone, respondent and another person, or someone else alone.

Questions relevant to the time spent with the pet asked for the following information: how much of the time the respondent and pet were in the same room when the respondent was at home (categories of almost never or a little, some, and most or all of the time); hours per day spent out doors with pet; hours per day petting pet; hours per day talking to pet; and amount of time spent with pet compared to other people respondent knew with pets (categories of much less or a little less, about the same, and a little more or much more).

Affective attachment was measured by a single question with five response alternatives: "And, would you say your pet is: extremely important to you; very important to you; fairly important to you; or not at all important to you?" Last, respondents were asked about the benefits of owning a pet and the negative aspects of owning a pet. These were open-ended questions that were subsequently coded into categories by the investigators. The benefit minus cost difference was determined by subtracting the number of negative categories mentioned from the number of positive categories.

RESULTS

Of the 1,036 respondents assessed at baseline, 938 (91%) remained in the final sample. Respondents were excluded if they missed more than one interview in Waves 2 through 4 or in Waves 5 through 7. Stated differently, respondents could miss a maximum of one interview in the first 6 months of the study and one in the second 6 months of the study and still be included in the final sample. An examination of the baseline characteristics showed that respondents in the final samples were younger and more likely to be employed than those who dropped out of the study. The two groups were comparable in presence of chronic health problems, gender composition, income, racial-ethnic composition, education, living arrangement, presence of pets in the household, social network involvement, depressive symptomatology, and the experience of recent life events.

The demographic characteristics of the respondents are presented in Table 1. For purposes of analyses, all demographic variable were collapsed and treated as dichotomous.

As can be seen, two thirds of the sample was between the ages 65 and 74 years, with one third of the respondents being 75 years and older. A greater proportion of the sample was female (60%) than male (40%), and the majority classified themselves as non-Hispanic White (89%). Half had incomes below $15,000, and two thirds had a high school education or less. Most (84%) were presently not working. Although more than half (56%) of the respondents currently were not married, the majority (58%) shared their households with someone. One third (37%) of the respondents had household pets, a proportion comparable to that in two other studies of the elderly (Garrity, et al., 1989; Ory & Goldberg, 1983), and more than half (57%) had one or more chronic health problems.

To examine the prospective relationship of baseline characteristics with utilization of services, the data on doctor contacts were aggregated across the six waves (Waves 2 through 7). Each respondent received a total doctor contact score for the study year and subscores of respondent-initiated contacts and physician-initiated contacts. Respondents who missed one interview in either Waves 2 through 4 (Months 2, 4, 6) or Waves 5 through 7 (Months, 8, 10, 12) or both were assigned for the wave they missed the average value of contacts in the other two waves in the 6-month period.

The three measures of doctor contacts were each regressed on the baseline measures. A hierarchical procedure was followed, with demographic variables (sex, age, race, education, income, employment status, social network score, and chronic health problems) entered on the first step and pet ownership entered on the second step. This analysis tested whether pet ownership accounted for a significant proportion of the variance in doctor contacts once the variance attributable to other demographic characteristics had been removed. After demographics and health status were controlled for, respondents with pets had fewer total doctor contacts ($\beta = -.07$, p .05) and respondent-initiated contacts ($\beta = -.07$, p .05) than those without pets. The two groups were comparable with regard to doctor-initiated contact over the 1-year period. Not surprisingly, the presence of chronic health problems was related to higher scores on all three contact measures, as was lower income. Men reported more respondent-initiated contacts than women. In sum, health status, income, and pet ownership were the major predictors of doctor contacts over a 1-year period.

The second step in the analysis was to determine whether the pet ownership might moderate the impact of psychological distress on doctor contacts. The previous regressions were repeated with the following revisions. Life events and depression at baseline were included as predictor variables, and both the Life Event X Pet Ownership interaction in term were entered on a third step in the regression analysis. With regard to total doctor contacts, poor health (β = .19, p .0001), lower income (β = -.11, p .01), the experience of greater number of life events in the previous 6 months (β = .15, p .001) and the Pet Ownership X Life Events interaction term (β = -.11, p .05) each made a significant independent contribution to the regression equation. The full model is presented in Table 2. To clarify the direction of the interaction effect, the mean doctor contacts were calculated within the four cells of the pet ownership by life events (split at the median) cross-classification). The pattern of means and tests of simple main effects support the hypothesis that pet ownership moderates the impact of life events on doctor contacts. Specifically, for respondents without a pet, the experience of many compared to few life events in the 6 months prior to baseline resulted in significantly more total doctor contacts during the study year (10.37 vs. 8.38, p .005). Life events were unrelated to doctor visits among respondents with a pet (8.91 contacts for those with many life events and 7.90 for those with few life events, ns).

Parallel findings emerged for the two subscores of doctor contacts. For respondent-initiated contacts, being male (β = -.11, p .01), being in poor health (β = .17, p .001), and having more life events (β = .13, p .01) each independently predicted contacts. Poorer health (β = .14, p .001) and many life events (β = .09, p .05) predicted doctor-initiated contacts. Although Pet Ownership X Life Events interaction was not statistically significant (β = -.08, p .13 for respondent-initiated contacts; β

= .09, p .07 for doctor-initiated contacts), the pattern of means supported the finding that doctor contacts increased as life events accumulated for nonowners, but not for pet owners.

The final set of regressions classified pet owners by type of pet. Thus, the pet ownership variable in the first set of analyses was cat owners (n = 141) compared with nonowners, then dog owners (n = 202) compared with nonowners, and bird owners (n = 45) compared with nonowners. There were too few fish owners or owners of other pets to be analyzed separately. (It should be noted that the sum of these subgroups totals more than the number of respondents because some respondents with pets had more than one pet.) These analyses showed that the Life Events X Pet Ownership interaction term was a significant predictor of doctor contacts (total, p .05, and respondent-initiated, p .05) for dog owners, but not for either cat or bird owners. Specifically, for respondents not owning a dog, doctor contacts increased as life events increased (10.39 compared to 8.37, p .01, for high and low life events, respectively). Total doctor contacts for the study year were 8.62 and 7.75 (ns) for dog owners reporting high and low life events, respectively. With regard to respondent-initiated contacts, respondents not owning a dog and reporting many life events had more contacts than those with fewer life events (5.14 compared to 3.96, p .001). Among dog owners, life events were unrelated to respondent-initiated doctor contacts (3.73 and 3.77 ns).

To explore the stress reduction aspects of dog ownership, dog owners (n = 201) were compared with pet owners who did not have a dog (n = 110) concerning their reported relationship with their pet. (Note that these analyses were for 201 rather than 202 dog owners because 1 dog owner answered the questions with regard to another type of "favorite" pet.) First, concerning time spent with their pets, dog owners relative to owners of other pets spent more time outdoors with their pets (1.43 hr per day vs 0.59 hr per day), t(282) = 4.96, p.001; spent more time talking to their pets than other people they knew with pets t(297) = 3.26, p.001. Dog owners felt more attached to their pets, t(309) = 3.30 = 2.47, p.01, than did owners of other pets. Furthermore, analyses of the specific positive and negative aspects of pet ownership indicated that dog owners were more likely than owners of pets other than dogs to mention that their pets make them feel secure, x2(1, N = 307) = 51.67, p.0001, and slightly more likely to mention that their pets provide love x2(1, N = 307) = 3.13, p.08. Owners of pets other than dogs were more likely to mention that their pets provide cheer or entertainment, x2(n = 307) = 4.97, p.05. No differences as a function of type of pet emerged in the frequency of citing specific negative aspects of pet ownership.

Table 1

Demographic and Health Characteristics of Medicare Enrollees in an HMO

Variable	%	n
Sex		
Male	40	379
Female	60	559
Age		
65 to 74	69	650
75 and older	31	288
Racial-ethnic group		
Non-Hispanic White	89	836
Other	11	102
Income before taxes		
Less than $15,000	50	466
Greater than $15,000	39	368
Missing Data	11	104
Education		
12 years or less	62	578
13 years or more	38	360
Marital Status		
Not married	56	522
Married	44	416
Employment Status		
Not employed	84	786
Employed	16	152
Living arrangement		
Lives alone	41	383
Lives with others	58	545
Missing data	1	10
Pet ownership		
No pet in household	63	593
Pet in household	37	345
Chronic health conditions		
None	43	402
One or more	57	536

Note: HMO = Health maintenance organization, N = 938

This 1-year prospective study suggests that pet ownership influenced the physician utilization behavior of the elderly. When sex, age, race, education, income, employment status, social network involvement, and chronic health problems were controlled for, respondents with pets reported fewer doctor contacts during the year than those without pets. This effect was particularly pronounced for respondent-initiated doctor contacts, suggesting that discretionary contacts were influenced more than physician-initiated contacts. Furthermore, pets seemed to help their owners in times of stress. The accumulation of stressful life events was associated with increased doctor contacts for respondents without pets; however, this relationship did not emerge for pet owners. Again, these analyses controlled for respondents' health status, depressed mood, and other demographic characteristics. Depressed mood was not itself a predictor of doctor contacts, nor did it interact with pet ownership. Additional analyses indicated that the physician utilization behavior of dog owners alone was unaffected by the accumulation of stressful life events.

An examination of the specific stressful events that were endorsed by this elderly population showed that those occurring most frequently were loss events. About one quarter of the sample (26%) had experienced the death of a close friend in the 6 months preceding the baseline interviews. The death of a close family member and major illness of respondent's spouse were each reported by 13% of the sample, and the remaining seven events were endorsed by less than 10% of the respondents. Thus, a recent loss of companionship was common.

Respondent-generated benefits of pet ownership (open-ended question) indicated that fully three quarters of the pet owners mentioned that their pets provided them with companionship or company. Feelings of security (25%) and feeling loved (21%) were the next most frequently cited benefits. Taken together, these data suggest that life events may be arousing needs for companionship that in turn may result in doctor contacts. This may occur because either doctor contacts satisfy the desire for companionship or the companionship loss is exacerbating other health concerns. For pet owners, however, it seems that their companionship needs are met partially by their pets. Therefore, pet owners do not show an increase in physician utilization with increasing life events.

Regarding type of pet, the data showed that owning a dog provided a stress buffer, whereas owning other types of pets did not. Apparently, dog owners have a qualitatively different relationship with their pets than do owners of other pets. Dog owners reported spending more time outdoors and talking with their pets than other pet owners in this sample and felt that, in comparison to others they knew with pets, they spent more time with their pets overall. Both talking and time outdoors have clear companionship functions. In addition, spending time outdoors may be either a contributor to or a consequence of increased physical or mental vigor, which in turn could be related to physician contacts. The analyses controlled for chronic health problems and depressed mood, however, which suggest that the benefits of dog ownership are not mediated solely via the greater activity level of the owners. Although the presence of chronic health problems is a crude measure of health status, these findings were replicated when another measure of health (self-rated health status) was substituted for chronic health problems.

Probably of greater importance than the data indicating that dog owners felt more attached to their pets than did other owners. Two studies of the elderly found that greater attachment to one's pet was associated with better mental health (Garrity, et al., 1989; Ory & Goldberg, 1983), and attachment was associated with better health when human companionship was inadequate (Garrity, et al., 1989). Moreover, a study of elderly persons yielded reports of greater pet involvement if the pet was a dog than a cat (Lago, Knight & Connell, 1983). Also, 50% of the dog owners said they spent 24 hr a day with their dogs, compared to 7% of the cat owners (Lago, et al., 1983). Finally, dog owners in the current study felt that the benefits of owning a pet outweighed the cost to a greater degree than did owners of other pets. Particularly salient among the benefits was security -- provided much more by dogs than other pets. A sense of security may be especially important to the urban elderly who constituted our sample.

Altogether, these data indicate that owning a pet, particularly a dog, may reduce the demand for physician services among the elderly. As all analyses controlled for health status, it appears that pet ownership is primarily influencing social and psychological processes rather than physical health. Indeed, records of physician utilization behavior are thought to reflect the individual's social as well as medical history. Further support for this notion comes from data indicating that pet ownership reduces demand for care in time of stress. This latter finding is consistent with the growing literature on the role of social support in buffering the potentially negative consequences of stressful life events (Cohen & Wills, 1985; Kessler & McLeod, 1985). It has been observed that only those social relationships that provide appropriate forms of support can act as effective buffers (Cohen & McKay, 1984). Accordingly, dogs more than other pets provided their owners with an object of attachment.

Table 2

Demographic Variables, Stressful Life Events, and Pet Ownership as Predictors of Utilization of Physician Services

Variable	B	SE	P
Sex (1=female)	-.07	.58	—
1=Non-Hispanic White)	.00	.07	—
Chronic health problems (1=1 or more)	.19	.57	.0001
Employment Status (1 = employed)	-.04	.75	—
Education (1 = high school graduate and beyond)	-.03	.57	—
Social network involvement (possible range from 0 to 49)	.07	.03	—
Income (1 = $15,000 or greater)	-.11	.59	.01
Life Events (possible range from 0 to 10)	.15	.37	.001
Pet ownership (1 = pet)	-.03	.84	—
Depression (possible range from 0 to 60)	.08	.05	.05
Life Events X Pet ownership	-.11	.63	—
Depression X Pet ownership	.05	.07	—

References

Amster, L.E. & Krauss, H.H. (1974) The relationship between life crisis and mental deterioration in old age. International Journal of Aging and Human Development, 5, 51-55.

Barsky, A.J., Wyshak, G., & Klerman, G.L. (1986). Medical and psychiatric determinants of outpatient medical utilization. Medical Care 24, 548-560.

Berkman, L.F., Berkman, C.S., Kasl, S., Freeman, D.H. Jr., Leo, L., Ostfeld, A.M., Cornoni-Huntley, J., & Brody, J.A. (1986) Depressive symptoms in relation to physical health and functioning in the elderly. American Journal of Epidemiology, 124, 372-388.

Brickel, C.M. (1986) The therapeutic roles of cat mascots with a hospital-based geriatric population. Gerontologist, 19, 369-372.

Broadhead, W.E., Kaplan, B.H., James, S.A., Wagner, E.H., Schoenbach, V.I., Grimson, R., Heyden, S., Tibblin, G., & Gehlbach, S.H. (1983) The epidemiologic evidence for a relationship between social support and health. American Journal of Epidemiology, 117, 521-537.

Cohen, S., & McKay, G. (1984) Social support, stress, and the buffering hypothesis: A theoretical analysis. In A. Baum, J.E. Singer, & S.E. Taylor (Eds.) Handbook of psychology and health (Vol. 4, pp. 253-267). Hillsdale, NJ: Erlbaum.

Cohen, S. & Wills, T.A. (1985) Stress, social support, and the buffering hypothesis. Psychological Bulletin, 98, 310-357.

Comstock, G.W. & Helsing, K.J. (1976) Symptoms of depression in two communities. Psychological Medicine, 6, 551-553.

Frerichs, R.S., Aneshensel, C.S., & Clark, V.A. (1981) Symptoms of depression in Los Angeles County. American

Journal of Epidemiology, 13, 691-699.

Friedmann, E., Katcher, A.H., Lynch, J.J., & Thomas, S.A. (1980) Animal companions and one-year survival of patients after discharge from a coronary care unit. Public Health Reports, 95, 307-312.

Garrity, T.F., Stallones, L., Marx, M.B., & Johnson, T.P. (1989) Pet ownership and attachment as supportive factors in the health of the elderly. Anthozoos, 3, 35-44.

Husaini, B.A., Neff, J.A., Harrington, J.B., Hughes, M.D., M & Stone, R.H. (1980) Depression in rural communities: Validating the CES-D Scale. Journal of Community Psychology, 8, 20-27.

Kahana, E. & Kahana, B. (1983) Stress reactions. In M. Lewisohn & L. Teri (Eds.) Clinical geropsychology: New directions in assessment and treatment (pp. 139-169). New York: Pergamon Press.

Kessler, R.C. & McLeod, D.J. (1985) Social support and mental health in community samples, in S. Cohen & S.L. Syme (Eds.) Social support and health (pp.219-240) Orlando, FL: Academic Press.

Lago, D., Connell, C.M., & Knight, B. (1983) A companion animal program in M.A. Smyer & M. Gatz (Eds.) Mental health and aging (pp. 165-184). Beverly Hills, CA: Sage.

Lago, D.J., Knight, B., & Connell, C. (1983) Relationships with companion animals among the rural elderly. In A.H. Katcher & A.M. Beck (Eds.) New perspectives in our lives with companion animals (303-317), Philadelphia, PA: University of Pennsylvania Press.

Lawton, M.P., Moss, M. & Moles, E. (1984) Pet ownership: A research note. Gerontologist, 24, 208-210.

Lubben, J.E. (1984) Health and psychosocial assessment instruments for community-based long-term care: The California Multipurpose Project experience. Unpublished doctoral dissertation, University of California, Berkeley.

Lubben, J.E. (1988) Assessing social networks among elderly populations. Journal of Family and Community Health, 11, 42-52.

Mechanic, D. (1972) Social psychologic factor affecting the presentation of bodily complaints. New England Journal of Medicine, 286, 1132-1139.

Mugford, R.A. & M'Comisky, J.G. (1975) Some recent work on the psychotherapeutic value of cage birds with old people. In R.S. Anderson (Ed.) Pet animals and society (pp. 54-65). London: Bailliere Tindall.

Murrell, S.A., Himmelfarb, S., & Wright, K. (1983) Prevalence of depression and its correlates in older adults. American Journal of Epidemiology, 117, 173-185)

Muschel, I.J. (1984) Pet therapy with terminal cancer patients. Social Casework, 451-458.

Ory, M. & Goldberg, E. (1983) Pet ownership and life satisfaction in elderly women. In A.H. Katcher & A. Beck (Eds.) New perspectives on our lives with companion animals (pp. 803-817). Philadelphia, PA: University of Pennsylvania Press.

Perlman, D. & Peplau. L.A. (1984) Loneliness research: A survey of empirical findings. In L.A. Peplau & S.E. Goldston (Eds.) Preventing the harmful consequences of severe and persistent loneliness (DHHS Publication No. ADM 84-1312, pp. 13-46). Rockville, MD: National Institute of Mental Health.

Radloff, L.S. (1977) The CES-D Scale: A self-report depression scale for research in the general population. Applied Psychological Measurement, 1, 385-401.

Rahe, R.H. & Arthur, R.J. (1978) Life change and illness studies: Past history and future directions. Journal of Human Stress, 4, 3-15.

Regier, D.A., Goldberg, I.D., & Taube, C.A. (1979) The de facto U.S. mental health services: A public health perspective. Archives of General Psychiatry, 35, 685-693.

Robb, S.S. & Stegman, C.E. (1983) Companion animals and elderly people: A challenge for evaluators of social support. Gerontologist, 23, 277-282.

Shuval, J. (1970) The social functions of medical practice. San Francisco: Jossey-Bass.

Soares, C.J. (1985) The companion animal in the context of the family system. Marriage and Family Review, 8, 49-62.

Tessler, R., Mechanic, D., & Dimond, M. (1976) The effect of psychological distress in physician utilization: A prospective study. Journal of Health and Social Behavior, 17, 353-364.

Waxman, H.M., Carner, E.A., & Blum, A. (1982) Depressive symptoms and health service utilization among the community elderly. Journal of the American Geriatric Society, 31, 417.

Weissman, M.M., Sholomskas, D., Pottenger, M., Prusoff, B.A., & Locke, B.Z. (1977) Assessing depressive symptoms in five psychiatric populations: A validation study. American Journal of Epidemiology, 106, 203-214.

The Journal of Personality and Social Psychology, 1990, Vol. 58, No. 6, pp. 1081-1086. Copyright 1990, by the American Psychological Association. Reprinted by permission.

The Latham Letter, Vol. XII, No. 2, Spring, 1991, pp. 1, 16-21.

GRIEF

When The Bond Is Broken: Companion Animal Death And Adult Human Grief

Mary M. Bloom, M.A.

A 124 page thesis was presented to the faculty of California State University at Dominguez in candidacy for a Master of Arts degree in Sociology by Mary M. Bloom in the Summer of 1986. Its title: "When the Bond is Broken: Companion Animal Death and Adult Human Grief."

Mary Bloom's impressive document reflects extensive research and, in our opinion, is responsible and contributes new knowledge to the field of "pet" bereavement. In addition to extensive statistical reporting, the publication considers significant problems about companion animal loss.

The Latham Letter is pleased to have permission to publish the "Summary and Conclusions" of Mary Bloom's thesis.

Results of this study have confirmed some important aspects of previous work done in the area of the human-animal bond. At the same time, other findings have added to and, in some cases, differed from pre-existing data.

The essays discussed in Chapter Eight substantiate a view that the loss of an animal companion has the potential to create grief responses in adults which paralleled those experienced with the loss of human companions. Lending further support to this view are the data collected on bereavement difficulties. Over three-quarters of those interviewed experienced difficulty or disruption in their daily routine immediately following their "pet's"* death. Most experienced classic symptoms of depression such as sleeping and eating difficulties and decreased social activity. In addition, over one-third of those experiencing problems said they took time off work after their "pet's" death and another one-third experienced difficulties in their relationships with other people. These results are similar to those previously obtained by Quackenbush and Glickman (1983).

Analysis of the data in degree of attachment as determined by the attachment index, provided some interesting and sometimes unexpected results. A high proportion of those sampled (63%) were highly attached to the deceased animal companion. The crosstabulations of degree of attachment by the demographic characteristics of the sample revealed that those between the ages of 40 and 49 were less likely than those in the other age categories to be highly attached. A possible reason for this result is discussed below in relation to the findings on number of children in residence at "pet"'s"death. The elderly over age 60) were only slightly more inclined than those in the remaining age groups to be highly attached. The results, therefore, do not provide direct evidence to suggest that degree of attachment is positively related to age within the sample.

Female respondents were substantially more likely than males to score highly on the attachment index. Of the 73 females on the sample, almost three-quarters were highly attached compared to only 41% of the 27 males. Again, this result may be explained by subsequent findings on attachment. For example, marital status and other home-related relationships were found to be determiners of degree of attachment. Generally speaking, respondents who were not married, lived alone or with just one other person, and had no children in the household were most likely to be highly attached to the deceased animal companion.

The results seem to indicate that fewer home-based human relationships are associated with stronger attachment. The data support the hypothesis that as human family size decreases, the social role and status of animal companion increases in importance. Females were less likely than males to be married and more likely to be separated or divorced. Also, the vast majority (83%) of those who lived alone were female, and females were slightly less likely than males to have children living with them at the time of their "pet"'s"death. All of these factors may contribute to the high degree of attachment found among female respondents and one, number of children in residence, may help to explain the relatively low degree of attachment among respondents in the 40 to 49 age group. This group was most likely to have one or more children in the household at "pet" death. Stronger attachment found among females may also be a reflection of traditional male/female roles. Quackenbush and Glickman (1983:385) suggest that it is generally more acceptable for females to show emotion or affection particularly for a "pet".

Only one characteristic relating to household size proved not to be significant in determining degree of attachment to the deceased animal companion. Surprisingly, degree of attachment among respondents with "multiple-"pet" households was as strong as it was among those with only the deceased animal companion living with them. It is speculated that positive attitudes toward animals in general which may contribute to having more than one "pet" may explain this finding.

The species (dog or cat) of the deceased animal companion also showed not significant in predicting degree of attachment. Instead, the length of time the respondent and animal companion lived together and other conditions relating to the quality of the human-animal relationship proved to be very important.

Those respondents whose "pet" lived with them six years or longer and whose "pet" resided inside the house or spent an equal amount of time inside and outside were most likely to be strongly attached. There was also a much stronger tendency for those who were primary caretakers and those who talked or touched their "pet" more than once a day to be highly attached. These results support the concept that the longer a person and animal companion live together and the more frequent the contact between them, the more likely it is that their relationship will grow stronger. The vast majority (85%) of females indicated that they were the primary caretaker of the deceased animal companion. This, no doubt, also provided females the opportunity to develop a stronger relationship with their "pet".

Further data confirm the observation of several authors that the tendency to experience bereavement difficulties following animal companion loss is positively related to degree of person to "pet" attachment. The results show that the more attached to the deceased animal companion the respondent was, the more likely he or she was to experience bereavement difficulties immediately following "pet" death. The crosstabulations of the question of bereavement difficulties by the individual characteristic of the respondent, the animal companion, and the human-animal relationship seem to reflect the findings on degree of attachment discussed above.

In general, there was no evidence to suggest that the tendency to experience difficulties after "pet" death increased with age within the sample. A higher percentage of those in the 40 to 49 age group indicated no difficulties whereas those in the remaining age groups were fairly evenly distributed.

Females indicated bereavement difficulties associated with "pet" death much more frequently than males, and the tendency to experience difficulties was stronger among respondents who were not married at the time of their "pet"'s"death.

More significant than marital status in predicting difficulties associated with animal companion bereavement, however, were factors relating to a household size and makeup. The tendency to experience difficulties was significantly stronger among respondents who lived alone or with just one other person and among those who had no children living in the household at "pet" death. The data support the hypothesis that as human family size decreases, animal companion loss becomes more stressful.

Neither the number of "pet"s in the household nor the species (dog or cat) of the deceased animal companion was a determining factor in predicting the likelihood of experiencing bereavement difficulties. These results differ from the research of Quackenbush and Glickman (1983) in which bereaved cat "owners" and those with only one "pet" were found to be more prone to a social work intervention following "pet" death.

In this research, the length of time the respondent lived with the animal companion was found to be a determinant of the likelihood of experiencing difficulties. There was a greater tendency for those whose animal companion lived with them for six years or longer to experience difficulties. Results tend to imply that the process of euthanasia may increase the likelihood of a stressful bereavement. Eighty-two percent of those who euthanized compared to 71% of those who did not experienced bereavement difficulties.

The variables relating to the quality of the human-animal relationship were significant in determining the tendency to experience difficulties. Those who indicated that they were primary caretakers and those who said they talked to or touched their "pet" more than once a day were substantially more likely to experience difficulties associated with the loss of their "pet". Also, the tendency to experience difficulties was much less among those whose "pet" resided outside of the house most of the time. These results probably reflect the findings on attachment discussed earlier.

The data on health problems associated with "pet" death

provide initial support for the view that animal companion death can adversely affect the health of a certain portion of bereaved "pet" owners." Forty-one percent of the respondents felt that the death of their "pet" had a negative effect upon their health. The majority (75%) said they felt anxious or depressed. Others said they experienced more frequent colds or flu, while some increased smoking or drinking or experienced weight loss. In regard to age, those in the 30 to 39 age group were most likely to experience these immediate or potential effects upon their health. Regarding gender, over one-half of the females compared to only 15% of the males said they felt negative health effects. As indicated earlier, females were substantially more likely to experience bereavement difficulties immediately following "pet" death. More long-term epidemiological research needs to be done to confirm these data.

The crosstabulated data on long-term effects suggest a strong tendency for those over 60 years of age to experience negative effects upon their sense of well being as a result of "pet" death. The elderly in the sample were much more likely to agree with statements (since the death of my "pet") "Loneliness is a greater problem," "I feel no one really needs me," "I worry more about my personal safety," or "Health problems have become more frequent." Those least likely to experience these effects were between 18 and 29, while those remaining age categories were fairly evenly distributed.

Perhaps due to losses associated with aging, the elderly are more vulnerable to feeling the loss of those benefits provided by animal companionship which increase feelings of personal safety or self-esteem or provide increased opportunities for exercise and socialization. Also more likely to experience long-term effects were females, those in one-person households, and those with no children living with them.

No definite conclusions could be drawn from the data collected on the ethnicity or socio-economic status of the respondent. Due to the low number of respondents in ethnic categories other than Caucasian, the crosstabulations were eliminated from analysis. Similarly, only one respondent fell into the lower socio-economic category and was, therefore, eliminated from analysis. The results derived from the crosstabulation of bereavement difficulties by socio-economic status are, therefore, inconclusive.

CONCLUSION

From this research, it can be concluded that variables relating to gender, marital status, household size and makeup, and the length and quality of the human-animal relationship may determine the likelihood of experiencing stressful bereavement following companion animal death.

The research findings suggest that female "pet" owners," those who live in one or two-person households, and those who do not have children living with them may be at an increased risk of problem bereavement. Those who are not married may also be more likely to experience difficulties associated with "pet" death.

The results of this research do not provide evidence to suggest that older adults or the elderly are more likely than younger adults who are strongly attached to their companion animal to experience bereavement difficulties immediately following "pet" death. The results do indicate, however, that possibly due to losses associated with aging, the elderly may be more vulnerable to experiencing more long-term effects of "pet" death such as increased feelings of loneliness or decreased feelings of personal safety or self-esteem. This finding is consistent with the observation of Westbrook (Los Angeles Times, 1983: Part V:5) that because of losses experienced in the past, the elderly may be better prepared to deal with the initial loss of a "pet" but may be more vulnerable to the loss of whatever benefits animal companionship may have provided them.

The results suggest that the longer a person and animal companion live together and the more frequent the contact between them, the more likely it is that bereavement difficulties will occur. The findings also provide some evidence to suggest that the process of euthanasia may increase the likelihood of experiencing bereavement problems.

It is hoped that this study will inspire others to further investigate the human-companion animal bond, the effect of its disruption, and those factors which may be related to bereavement difficulties. By identifying those who are most vulnerable to this kind of a loss, bereavement care and support could be provided in a preventive as well as crisis situation.

The Latham Letter, Vol. VIII, No.1., Winter 1986/87, pp.6-8.

Death with Dignity

The Implications of Human/Animal Bond Interactions Within the Hospice Programs

Thomas E. Catanzaro, D.V.M.

Jodie L. Sell

Hospice is a philosophy, a special kind of family-oriented health care program that provides services to patients who are terminally ill. It is a system of care that seeks to restore dignity and a sense of personal fulfillment to the dying. The focus is on the patient and the family, rather than on the disease, and the aim is to improve the quality of life that remains. Our recent survey has shown that 70 percent of the retired families consider the pet as a full family member and 99 percent of the families believe children should have pets; the implications of this human/animal bond involve the patient of any age, the family, and the hospice team.

THE HOSPICE CONCEPT

While there are various concepts of what the term "hospice" really means, for the purpose of this presentation **HOSPICE** is defined as:

"a program which provides palliative and supportive care for terminally ill patients and their families, either directly or on a consulting basis with the patient's physician or another community agency such as a visiting nurse association. Originally a medieval name for a way station for crusaders, where they could be replenished, refreshed, and cared for; it is used here to describe an organized program of care for people regardless of age going through life's last station. The whole family is considered the unit of care and this care extends through the mourning process. Emphasis is placed on symptom control and preparation for and support before and after death."[1]

The hospice movement in the United States is modeled after the British movement, begun in 1967. Hospice care is distinguished from other health care modalities by four basic principles: a) the unit of care is both the patient and the family, b) care is planned and coordinated by a team of individuals to insure that all needs are met, c) efforts are aimed at caring instead of curing, and d) hospice care for the family is continued during the bereavement period.[2] This family orientation means the hospice team must be very aware of the home environment, to include the role of animals.

The growing hospice movement (there are 1,800 hospices in the United States today) may be attributed to several factors. Rapid advances in technology and medical science make it possible to delay death almost indefinitely. Medical and nursing schools emphasize the curative, rather than the supportive. Our contemporary society has dictated that death is to be delayed at all costs. With these prevailing attitudes, emphasis is frequently geared toward quantity of care instead of quality of care; this is why most acute care hospitals are technology-intensive. However, 75 percent of all patients in these acute care hospitals will not be cured; they have chronic and degenerative diseases which will result in death.[3] Many of these patients are elderly. There are presently 26 million Americans older than 65, and the number is growing.[4] One-fifth of the nation's population is projected to be over 65 by the year 2050.[5] With advancing age comes an increase in chronic and debilitative diseases. Last year 26 million people died in the United States and over 70 percent of those were age 65 or older.[6]

Despite rapid advances in technology, there are times when health cannot be restored, and the emphasis or care must shift from curing a patient to caring for the patient without the ultimate goal of restoring health. However, we as a society have abdicated our role in death and dying to professionals. Death is viewed as a personal threat as well as a professional failure. Decreasing quantity of care for a patient is often equated with decreasing quality of care. In the case of the terminally ill patient, however, a significant amount of care is traditionally delivered whose benefits are probably not worth the cost. If quality of life rather than longevity is the goal, the dying patient may benefit more from alternative treatments in other settings which emphasize support and caring instead of curing. This, in essence, is the hospice philosophy: to create an environment in which the terminally ill patient can live and die with as much comfort and dignity as possible. Hospice attempts to deal with the needs of the whole person. This is the concept that makes the human/animal bond and the principles of animal facilitated therapy and animal facilitated communication so important in the hospice program.

ANIMALS IN HOSPICES

The use of animals in hospice programs can be evaluated from two viewpoints: the desires of the patient or the desires of the family. If the two viewpoints are compatible after interfamily negotiations, the hospice team can move to the next step, the selection of an appropriate animal. If the family and patient cannot reach a mutual agreement, the hospice team needs to educate the family on the expected benefits or reconsider the use of an animal as a treatment alternative.

The spectrum of animals available for hospice programs far exceeds the traditional problem of selecting which breed of cat or dog. Granted, the benefits of a purring cat or a tail-wagging dog come first to mind when "non-judgmental love" is discussed, but other companion animals are available. The aquarium provides a colorful, tranquil, low maintenance distraction, although the tactile interaction is missing. A vivarium is similar to the aquarium, is unusual and stimulates discussions, provides limited tactile stimulation if the patient does not fear reptiles or amphibians, and, like the aquarium, requires minimal care. Small caged mammals (mice, gerbils, hamsters, etc.) offer active, furry, tactile interactions that are entertaining, require low maintenance, and are generally economical. While urban-raised people can get some companionship and a limited opportunity to express love and affection, the small caged mammal often evokes rejection within rural families which consider rodents an occupational problem. Rabbits and guinea pigs, though still caged, offer a greater opportunity to some degree, are calm, and can become companions very similar to a cat or dog. The shorter life span of rabbits and guinea pigs has virtually no adverse impact on a hospice program due to the relatively short length of the health care encounter. Caged birds that are suitable for hospice programs are the psittacine (parakeet/parrot type) and finch-type species. These are colorful, vocal and entertaining. The psittacines are hand-tameable and can be taught to do tricks and communicate. If space or care is limited, a wild bird feeder placed near a window can provide a distraction, a topic for conversation, and often, an outlet for concern and caring.

The word "hospice" is not synonymous with a specific setting. It may take one of several forms: free-standing with hospital affiliation, a hospice unit within a hospital, or home care.[7] A hospice in a community may mean that volunteers are the driving force; or that a special ward of a hospital has been designated especially for the terminally ill; or a special staff in a separate building provides hospice care. Whatever the

setting, the concept remains the same. Attention is focused on the reduction of pain and relief of other symptoms to keep the patient as comfortable as possible, but expensive life-sustaining machinery is avoided.

THE VALUES OF HUMAN/ANIMAL BOND THERAPY WITHIN HOSPICE CARE PROGRAMS

In discussing values and economics within health care, we must keep three things in mind: a) resources are scarce in relation to human desires and wants, b) resources have alternative uses, and c) individual wants and levels of utility are highly subjective.[8]

One of the primary goals of economics is to determine the best way to allocate society's scarce resources. The optimal allocation of resources between two products depends upon the demand for the products and upon the costs of producing them.[9] The price system has generally been successful in offering the incentives to shift resources from one use to another as consumer tastes, technology, and populations change. This shift to more productive uses leads to a more efficient allocation of resources. However, in many ways the health care industry does not fit the conditions of a competitive economic system. Feldstein offers the following statement to describe the ideal economic criteria for evaluating medical system performance:

"A medical system which values economic efficiency in consumption and production will base its choice of the amount to spend on medical services on the criteria of satisfying consumer preferences; it will base its method of providing services on the criteria of least cost; and it will base its choice of the amount and method of medical services redistribution on the criteria of consumer preference.[10]"

Obvious economic considerations were evaluated in our survey by asking how much owners spent annually, per pet, for other than pet food. While the elderly/retired portion (n=156) of the survey population (n=961) showed that 10% reported $200 or more per year, 26% stated they spent between 0-$30/year, 29% spent $40-$70/year, 22% spent $80-$110/year, and 12% spent $120-$190/year, the general population spent a little less than this distribution.

The consumer preference for animal companionship has been repeatedly validated in recent value surveys. The most recent large-scale survey was conducted by Psychology Today, August 1984,[11] and it showed 99 percent of the owners talked to their pets, over half reported that their pets sleep in the same bed as a family member, and 75 percent reported that a pet made for more fun and laughter in the family.

The survey we conducted allowed us to evaluate 156 retired (over 50 years of age) military families, identify their specific values toward their pets, and compare them to the survey population in total (n=961). In comparing the results of the older retired family to the total populations, some trends become evident. It was reported that 24.5% usually celebrate their pet's birthday, (about 10% less than the general population surveyed) and when asked about display of their pet's picture, 58% displayed pictures in their home. These rates reflect more pictures but less birthday celebration than the total population surveyed, an indication of increased pet importance with less desire to "count the days." In reviewing the names that the elderly respondents gave their pets, a subjective review of the surveys reflected 48% of the pets had "people-type" names, 31% had been named after "things or physical traits," and 21% were given "animal-type" names. When compared to the total population of respondents, which reported that 23.4% were named after physical characteristics, 11.7% were named after TV show/movie/cartoon/or book characters, 9.0% were named after a person and 7.7% were named after a previous pet, the "belonging" status of the pet became far more evident. This was reinforced when the elderly were asked about reasons for selecting their companion animals; they rated, in order of priorities, the reasons as follows: pleasure (36%), companionship (35%), and protection (10%).

Many traits or attributes are credited to the companion animal during daily conversations, so the survey asked specifically what special characteristics the pet displayed with the family. The responses were placed on a line scale from "Great" through "Some" to "None," with the following results noted.

TOTAL POPULATION (N=961) GREAT DISPLAY OF TRAIT	CHARACTERISTIC	ELDERLY POPULATION (N=156) GREAT DISPLAY OF TRAIT
89.3%	Greets you upon coming home	90.6%
76.9%	Pet understands when you talk to him/her	84.4%
72.9%	Communicates to you	78.1%
59.6%	Demands for attention	55.4%
59.2%	Understands/sensitive to your moods	62.9%
49.7%	Stays close when you're anxious/upset	49.5%
44.9%	Sleeps with family member	32.3%
22.5%	Mimics your emotions	20.7%
11.8%	Hides or withdraws when you are anxious/upset	5.1%
10.6%	Expresses feelings that you cannot/do not	16.1%
3.8%	Develops illness when family tension high	5.1%

When asked directly to evaluate how important their pet was to their life, 50% reported "extremely important," 34% reported "very important," 14% reported "important," and 1.9% reported moderate to no importance. To better evaluate the importance of the companion animal, respondents were asked to rate the importance of their pet to them, in specific situations on a line scale from "great" to "some" to "none," with the following results:

SITUATION	TOTAL POPULATION		ELDERLY POPULATION	
	GREAT	NONE	GREAT	NONE
At all times	75.4%	1.7%	79.6%	0.0%
Temporary absence of spouse	73.1%	6.7%	70.6%	5.9%
Free time/relaxation	71.5%	2.8%	78.7%	0.0%
Childhood period	69.6%	9.7%	59.0%	19.7%
Sad, lonely, depressed	68.4%	5.1%	72.3%	4.0%
Marriage without children	58.6%	25.5%	58.5%	24.5%
Temporary absence of children	53.2%	17.5%	51.6%	18.8%
During illness/after death of other	52.0%	13.2%	60.0%	8.2%
During crisis/separation/divorce	50.3%	16.3%	53.5%	18.3%
During moves or relocations	48.2%	17.5%	44.1%	20.3%
Teenage period	44.4%	16.1%	47.5%	18.6%
Unemployment	35.6%	31.4%	34.0%	37.7%

It was interesting to note that while the intensity of feelings concerning the importance of the companion animal was often more polarized in the elderly than the population in general, the significance of death of the animal to the elderly was over 20% less (73.1% vs. 94.4% important to extreme loss). This is thought to be due to the greater acceptance of the death process by the elderly.

Besides questions concerning the importance of the pet to specific situations, other questions were posed to evaluate the respondents anthropomorphic tendencies. When asked how the companion animal fit into the family group, 71.9% stated that the pet had full family member status, 23.9% had friend status, and 3.2% of the pets were considered possession/owned property. While these responses reflected a greater value to the elderly retired population, 99 percent of the families surveyed felt children should have pets. Later in the survey, the question was rephrased and the respondent was asked if the pet was afforded "people" status in the family. A line scale was used and the respondents were asked to rate the status from "always" through "usually" and "sometimes" to "never." The survey indicated 78.4% of the elderly felt their pet was usually to always afforded "people" status, while only 2.6% stated they never gave their pet "people" status; this rate was about 10% higher than for the total surveyed population.

Utilization of companion animals to reduce feelings of loneliness, depression, or boredom has been well documented in the recent literature. [12,13] Individuals have also exhibited dramatic improvements in their ability to interact and communicate with other residents and staff. These behavioral effects, whether subjectively or objectively proven, have resulted in a decreased staff workload, as well as an improved cost-benefit ratio. This animal facilitated communication benefit could assist the hospice team at any point in the loss phenomenon shown below.

Kubler-Ross Loss Phenomenon in Death/Dying Bereavement (Stress Mediated)

DENIAL
    ANGER
        GUILT
            DEPRESSION
                ACCEPTANCE

The rising costs of health care, especially hospital services, have been common concerns in the health care literature. Nationwide, health care costs rose from $41.7 billion in 1965 to $286 billion in 1981. Hospital charges come to approximately $150 billion each year. Between 1970 and 1980 the cost of an average hospital stay soared from $670 to over $3,000.[14] However, resources available for health care have become increasingly constrained. Health care must be considered a commodity. In order to achieve economic stability in health care, it will be necessary to introduce competition by offering low-cost alternatives to traditional inpatient treatment.[15] In the case of the terminally ill patient, there must be accompanying shift of attitudes whereby quality of life rather than longevity is the therapeutic goal, and death is accepted as an integral and inevitable counterpart to living. By offering services to the chronic and terminal patient in a more personal, as well as cost-effective, manner, while maintaining quality care, the hospice serves as one such alternative in the production and delivery of medical services.[16,17]

THE FUTURE FOR HOSPICE PROGRAMS USING ANIMAL FACILITATED THERAPY

The proportion of elderly in the nation's population is rapidly increasing; the challenge of meeting the long-term care needs of the elderly over the next twenty to fifty years is enormous. It is apparent that this is the age group which will have the greatest impact on health planning, health care provision, and health care costs over the next several decades. We must come to the realization that resources available to meet health care needs are limited, and these resource constraints will make stricter allocation decisions inevitable. This is especially evident in the care of the terminally ill elderly. Although technology has brought great changes in the delivery of health care; not all persons will be able to benefit from this contemporary medical technology. This factor is especially true with the dying child, where "quantity" care is often detrimental to the child's mental health, as well as their personal rights as a patient.

The hospice approach is offered as one alternative in the production and delivery of medical services to the terminally ill regardless of age. Attention is focused on palliative care rather than curative, offering death with dignity. In addition, several elements of care usually not found in the acute care system are part of the hospice; family and patient as one unit, continuum of care available, symptom management and pain control, use of the interdisciplinary team, utilization of therapy extenders or facilitators, and bereavement counseling. Many of the benefits are difficult to measure, such as the effects of using the companion animal. While the literature is full of articles [18,19,20] about the role of animals with the elderly or with children, death is seldom mentioned.

TABLE 1

Potential Roles of Animals in the Grief Process

Companionship
Nonjudgmental Love
Security/Safety
Neutral Communication Point
Stress Reduction
Triangulation (3rd party role)
Reality Anchor during Reminiscence
Potential Distractions (eg: exercise)
Intangible Distractions (eg: mood)
Mandatory Distractions (eg: feeding/care)
Stability of Environment

It has been well documented that the elderly with companion animals perceive less loneliness and less emotional isolation, as well as being provided something to care for, to keep them busy, to watch in idle times, or to provide a stimulus for exercise. The importance of the companion animal in life review, or reminiscence, cannot be understated in the hospice program; the animal provides a sense of security, as well as a dependence, that often anchors the hospice patient to the realities of daily life.

The hospice movement is growing. It will play a major role in cost-containment and efficiency of health care delivery in the future, by offering better ways to care for the dying. The use of companion animals will aid in cost containment while increasing the humanity of the hospice team encounter.

SUMMARY

As explained earlier, throughout history a hospice has been a place where hospitality is given. In the twentieth century, hospice is still a program of hospitality; yet it is a program specifically designed to provide medical, emotional and practical care to individuals in advanced stages of disease. The adjunctive therapy that a companion animal provides cannot be matched in conventional medicine; our research shows that the role of the companion animal in giving and taking of nonjudgmental love can greatly effect the psychosocial posture of the family, and thereby the entire program.

Hospice programs have been labeled alternatives to euthanasia, death with dignity, a palliative care program for the patient and family, or simply, an alternative to acute care hospitalization. In fact, a hospice program is designed to provide a source of warmth, compassion, and dignity to the dying patient. Advocacy of the hospice program concept has been growing rapidly during the past decade; there are currently over 1,800 separate programs in the United States, most of them being home care. As this is extremely personalized medical and psychosocial care program becomes easier for the patient to access, the implications of animal facilitated therapy programs will become an essential consideration for the interdisciplinary hospice team's approach to patient and family care. Bereavement counseling for the hospice family continues beyond the life of the patient. Our survey shows that hospice teams and support groups should find the human/animal bond an important factor to utilize in assisting family members during the grief process (see Table 2).

TABLE 2

GRIEF IS A PROCESS, NOT A SINGLE FEELING; A PROCESS OF "LETTING GO" WITHIN THE LIFE PROCESS

ANTICIPATORY GRIEF STAGE	CRISIS GRIEF STAGE	CRUCIBLE GRIEF STAGE	RECONSTRUCTION STAGE
Death is expected	*Death occurs*	*After the funeral*	*Return to self senses*
Emotions:	**Emotions:**	**Emotions:**	**Emotions:**
Denial with hope	Shock	Pain and fear	Orientation to present
Hope with long range spiritual plan	Numbness	Blame and anger	New interests
Anger and/or guilt	Disorientation	Guilt	Self-growth
Withdrawal and social death as rehearsal	Disbelief	Reminiscence	**GETTING STUCK SIGNS:**
Over-compensations become smothering		Need to deal with emotional realities	2 weeks of insomnia
		Develop new roles for family members	Increased weight loss
			Increased booze
			Increase in destructive behavior
KEY PLAYERS:	**KEY PLAYERS:**	**KEY PLAYERS:**	**KEY PLAYERS:**
Family	Health care provider	Social counselors	The person himself
Ministry	Funeral director	Ministry	
		Significant others	

A hospice program is people--the patient, the family, and the staff; but it is also the community it serves. Each hospice program needs the support of the community and of informed volunteers. It also needs the skills of professionals from all of the health care disciplines. Human/animal bond implications are substantial, but they are still only adjunctive therapy factors. The "caring heart" pet must be utilized by "caring heart" health care providers. The human/animal bond implications for the health care delivery team still have many horizons to conquer.

References

1. A Discursive Dictionary of Health Care (1976), s.v. "Hospice."

2. Henry R. Rolka, "Quality Assurance for Terminally Ill." Hospital and Health Services Administration 28 (March/April 1983): p.72.

3. David Skeleton, "The Hospice Movement: A Human Approach to Palliative Care." Canadian Medical Association 126 (1 March 1982): p. 556.

4. Gregory F. Pierce. "The Elderly: What is to Be Done?" The Lutheran Witness. (October 1983): p 20.

5. R. William Warren. "Meeting the Needs of the Elderly Patient," Texas Hospitals 37 (January 1982): p. 8.

6. Lawrence Maloney, "A New Understanding about Death," U.S. News and World Report, 11 (July 1983): p. 62.

7. Barbara A. Johnson and Howard L. Smith. "A Strategy for Integrating Hospice, Hospital Care." Hospital Progress 62 (November 1981): pp. 52-53.

8. Victor R. Fuchs. Who Shall Live? (New York: Basic

Books, 1974). p.4

9. Edwin Mansfield. Economics: Principles, Problems, and Policies (New York: McGraw-Hill, 1975): pp. 44-45.

10. Paul J. Feldstein. Health Care Economics (New York: John Wiley & Sons, 1979): p. 11

11. Jack C. Horn and Jeff Meer. "The Pleasure of Their Company" Psychology Today (August 1984): pp 52-58.

12. The Pet Connection. CENSHARE, University of Minnesota, Minneapolis, MN 55455.

13. New Perspectives on Our Lives with Companion Animals. University of Pennsylvania Press, Philadelphia, PA.

14. Steven Findlay, "The Trend Toward Treatment at Home." Fact (July/August 1983): p. 30.

15. Orrin G. Hatch, "Home Health Care: Necessary Option for Older Americans," Hospital Progress 64 (April 1983): p.6.

16. "Medicare Covers Hospice Under New Federal Law" Hospitals 56 (1 December 1982): p.93.

17. Deborah Shapely, "Hospices Gain." Nature 2 (September 1982): p. 6.

18. L.K. Bustad. Animals, Aging, and the Aged. University of Minnesota Press, Minneapolis, MN.

19. S.A. and E. O'L. Corson. Ethology and Nonverbal Communications in Mental Health (Pergamon Press, 1980, Oxford).

20. B.M. Fogle. Interrelations Between People and Pets (Charles C. Thomas, Springfield, IL, 1981).

Thomas E. Catanzaro is a Doctor of Veterinary Medicine and Holds a master's degree in Health Care Administration from Baylor University, Currently he is a veterinary management consultant.

Jodie L. Sell B. Sc.N, M.H.C.A., Lt. Col., USAF, NC, is currently stationed in the Office of the Surgeon General, Health Facilities Division at Brooks Air Force Base in Texas.

The Latham Letter, Vol. VII, No. 4, Fall 1987: pp. 14-17.

Pet Loss Considered From the Veterinary Perspective

Eddie Garcia, D.V.M.

Because pets have a relatively short life span and sometimes find themselves in dangerous or defenseless situations, most of our clients will sooner or later experience the loss of an animal. In our practice, we encounter 20 to 30 pet deaths per month due to euthanasia, natural causes, or accidents. When pet loss occurs, it is my observation that about half of all clients openly display their grief. Clearly, practitioners who can deal effectively with human reactions to pet loss greatly enhance their professionalism and client relationships. Because nearly two-thirds of those who experience the death of a pet will replace their animals within a year,[1] the veterinarian's response to their loss also becomes a vital factor in their decision to retain the services of the same practitioner.

INTENSITY OF PET BONDING

One thing I have come to fully appreciate only recently in my 18 years of practice is the intensity of the bond between people and their pets. Believe me, we should never underestimate the depth of that relationship, because it is often much,

much greater than we think it is. To illustrate this, I found these results of a recent survey to be revealing:

> 90% of pet owners said their pet was "extremely important" or "very important" to them.
>
> 88% said that stroking a pet helped them relax.
>
> 87% considered their pet a member of the family.
>
> 75% said their pets increase fun and laughter in their families.
>
> 50% kept pictures of their pets at home, in their billfold, or at work.
>
> 50% allowed their pets to sleep with a family member.
>
> Nearly half said that having a pet made them feel safer from crime.
>
> 25% celebrated their pets' birthdays.
>
> Owners ranked their pets right behind family and friends and ahead of work in importance.
>
> Pet owners said their animal give them more satisfying lives and made them less lonely.

Finally, this compelling and perhaps even tragic statistic was cited--79% of pet owners said that sometimes their pet was their closest companion. This is a statement I can accept. I see a surprising and growing number of pets treated as child substitutes, some by owners with several children living in the home.

This survey, one of many accounts now surfacing on the importance of human-pet bonding, has a lesson for those of us in practice. That is, we cannot always tell in the clinic how strong that bond may be, it is often more intense than we

suppose, and we should always assume that a client had a close relationship with a deceased pet. Furthermore, an individual who presents a pet at the veterinarian's clinic usually represents a larger family unit, often including children, whose lives are touched by their animal. When a pet dies or is terminally ill, it is the veterinarian's responsibility to present various options to clients and advise them on behalf of their entire family. It is a serious part of pet practice.

THE STAGES OF GRIEF

Let me share a few commonly accepted facts about the human phenomenon of death and grief:

Grief is a natural response to many loss experiences-- death, divorce, relocation, job loss, illness, or disability -- and should not be denied, ignored, or considered inappropriate.

Grief is more pronounced when death is unexpected, and less intense if it can be expressed before anticipated loss (preparatory grief).

Our society is not one that copes particularly well with dying because we often have a fear and non-acceptance of death (we are a "death-defying culture" in the words of one bereavement counselor[2]) compared with some societies where death is openly mourned.[1]

Grief has several emotional stages. It is helpful for medical professionals to understand what they are because it will help them develop strategies and techniques for dealing with clients as they progress through each stage.

Most clinical descriptions of grief conform more or less to the pattern described in 1969 by Dr. Elizabeth Kubler-Ross in her book *On Death and Dying*.[3] The grief experience can be consolidated into three stages -- denial and depression, pain and anger, and acceptance and resolution. The last stage is particularly important because only when grief is resolved can the client resume a normal life, and can a satisfactory doctor-client relationship be restored. Thus, resolution of grief should be the ultimate goal in any case of pet loss.

The first grief stage of denial and depression often comes at a sub-conscious level. As a result, the client may not respond openly to news of a pet's mortal condition. However, the veterinarian should assume the client is experiencing the emotions of denial and depression to some extent, and adopt some of the following strategies:

Give the client time to accept the facts, then follow up a short time later.

Provide confirmatory evidence of the animal's condition (diagnostic results, for example). Some clients have asked me to let them see or keep their pet's medical files. At first I thought this request meant they were changing veterinarians. However, many came back, and I realized that client awareness of a deceased pet's medical history helped them resolve their grief.

Avoid giving false hope. It makes eventual death harder to accept. Give clients an accurate prognosis, then assure them you are doing all you can. Do not make the assumption they know you are doing everything possible. They *want* to be reassured.

Allow clients to visit hospitalized, terminally ill pets in private.

Call owners of hospitalized, seriously ill pets once or twice a day. They appreciate knowing you care. Sometimes I have a technician make such calls, but I try to make at least one daily contact myself.

The second grief stage, expression of pain and anger, is the natural response to an inability to control events. This stage usually produces the most difficult situations for veterinary personnel because anger is often directed at the person who is handiest, frequently the practitioner or clinic staff.

Clients who get angriest are those who are loneliest and least secure. Your best strategy is not to take client anger or bitterness personally, avoid reacting or being judgmental, allow the client to ventilate, then leave the client alone until he or she regains composure.

RESOLVING GRIEF

The third grief stage can be recognized when fond memories replace grief, and appreciation replaces the sense of loss. The time required to resolve grief varies, and may take months or years depending on how close the bond was between owner and pet. One study has shown that the bereaved pet owner experiences a three to four month grieving period on the average. Of course, there are clients whose lives revolve around their pets. Their grief may be acute and perhaps exaggerated.

A lack of support and understanding prolongs the resolution of grief. Thus, the practitioner's best strategy at this stage is to be supportive. Listen to clients' concerns, then tell them their feelings are normal. Provide an objective viewpoint, relate similar client experiences, and perhaps refer the client to a professional counselor. In talking to the grieving client, simple and sincere expressions of concern are always best. Your presence and willingness to listen are signs in themselves that you care.

Pets should not be replaced until the client has resolved the loss of the former pet, because the new animal, consciously or subconsciously, will not be given full acceptance. Occasionally, I will hear a bereaved client say, "I'll never own another pet." My response often is, "I know there will never be another Gigi, but I think you can share your life with another pet. You have enough love to do that." I also tell them that replacing the pet when the time is right is an affirmation of a positive relationship with the animal that died.

If you get clients thinking about pet replacement and avoid having them make commitments never to replace a lost animal, then they will not say, "Well, I told Doctor I would never get another pet, and now if I do he'll think I didn't care that much about Gigi." Avoid letting clients paint themselves into a corner. Make it easy for them to continue enjoying pet companionship. A receptive attitude toward obtaining a new pet is a definite sign that loss of the former pet has been resolved.

WHEN EUTHANASIA IS NECESSARY

For years, I encouraged clients not to be present during euthanasia because I assumed they did not want to see their animal put down. I now know that the opposite is true. Recently, I contacted several of my good clients whose pets I had to euthanize in past years, and asked them if they would have preferred to be present. I was surprised when every one of them said more or less the same thing, "I never mentioned this to you, but I wish I could have been there. I want to be there if we have to do this again."

Most clients think their veterinarians do not want them to be present when their pets are euthanized, and so choose not to ask to witness the event. I strongly believe that the client should be given the option, and so I ask them, "Do you want to be present?" Some of them will ask, "What do you think I should do?" I tell them, "It's perfectly alright for you to be there if you want to be."

I know this can create some problems. The client may become emotional. Maybe you cannot find a vein when the euthanizing agent is administered. There may occasionally be an unpleasant physical response to the drug. The animal may be irritable or in pain and difficult to handle. I find that these concerns can be effectively dealt with by discussing them in advance with the client. (One tip of avoiding problems with venipuncture is to insert an indwelling IV catheter before bringing the pet and client together.)

All things considered, I believe it is appropriate and beneficial to give clients the opportunity to be with their pets when euthanasia is performed. It is very important to allow a client to have a few private moments with the pet beforehand, whether or not they wish to be present at death. Most tell me how much they appreciated these last minutes to say good-bye to their pet.

Most authorities agree that the burden of the decision to euthanize a pet should rest with the client, not the veterinarian. In fact, when euthanasia is decided upon, the veterinarian should obtain a written authorization statement signed by the client. The veterinarian should present all pertinent facts, options, and probabilities, but allow the client to make the final decision. Afterwards, however, in a condolence card or conversation I will tell the client, "*We* made the right decision." The "we" is important because some clients will be shouldering a certain amount of guilt for deciding to terminate a pet's life. It is reassuring to them that you participated in their decision-making process.

PET DISPOSAL AND CONDOLENCES

It is very helpful in resolving the loss of a close pet to conclude the relationship in a way that has a sense of finality or completeness. This is one reason why I do not object if the client wishes to be present during euthanasia. It is an event that signals the conclusion of the relationship between owner and pet. Another way of achieving that end is to encourage the client's family to conduct a short memorial service or, if possible, to bury the pet in a suitable place at home, or both. Sometimes I will be asked to attend, and I always try to do so.

I find that nearly half of my clients want to bury their pet at home. I usually present that as their first option. About 20% want to use a pet cemetery, despite its expense. The other options to discuss are cremation and group burial. If clients want me to handle pet internment, I assure them that it will be done in an aesthetically appropriate way and that unacceptable methods of disposal will not be used. If you handle disposal of the pet, the client may not ask what you did with the body (in part because they do not want to ask an embarrassing question), but I can assure you they are wondering. So, I tell them how the body will be handled, a policy I have found that my clients appreciate. If the client is undecided about the method of pet disposal, I suggest placing the body in a special freezer at the clinic until a decision can be reached. Again, clients appreciate this accommodating stance during a difficult time.

An expression of condolence to a bereaved client is not only a matter of courtesy but an important factor in continuing a professional relationship with that individual. Condolences can be made with a phone call, or in a note or card. I use either method depending upon the person involved, usually within four days after the pet's death, because that is the time when the client's sense of loss is greatest.

In expressing condolences, I try to be simple and sincere. I will tell a client that I am sorry for their loss. If euthanasia was involved, I may tell them that we made the right decision, that the animal will endure no more suffering. I may say how much I admired the caring attitude the client displayed toward the pet. Perhaps because of my Hispanic background, it is easy for me to touch or put my arm around someone else. I do that with a client if it is natural and appropriate under the circumstances. I try to avoid maudlin or overwrought expressions of sympathy, which generally come across as insincere. In the case of elderly clients who are taking the loss of their pet badly (most do), I may contact their adult children to encourage them to lend extra support to their parent.

In special cases, I may send a contribution to a worthy charity in the name of the client's pet. Suitable recipients include the AVMA, schools of veterinary medicine, zoos, the Latham Foundation, and the Morris Foundation, to name a few. Donations should be made in addition to a written or verbal condolence sent immediately to the client, because it may take several weeks for the donation recipient to send an acknowledgement.

One time I sent a donation on behalf of a client, who later

received an acknowledgement saying, "In memory of your beloved friend, Dr. Garcia." The client, misinterpreting the ambiguously worded message, thought I had died, and called the clinic to express concern. She was assured that rumors of my demise were greatly exaggerated! Obviously, expressions of sympathy can sometimes go awry. They need to be handled with care, discretion, and attention to detail.

TIPS FOR CLINIC STAFF

Whenever a pet dies, either in or outside the clinic, as a last administrative duty, I have the patient's file up-dated so that vaccination reminders are not sent (an embarrassing oversight!). All staff personnel are advised of the pet's death, so they are aware of the client's loss and can express their own condolences if the opportunity presents itself.

A recent survey showed that in 80% of veterinary practices, an animal health technician had an active role in interacting or consoling clients who had lost a pet.[4] Clearly, a team approach among the veterinarian's staff is advisable in handling client bereavement. Here are some simple and effective actions that technicians can take in dealing with bereaved clients:

Talk soothingly to a sick or terminal animal when moving it, especially in the presence of the owner.

Remember to use the client's name and the pet's name.

Assist in making follow-up calls, both to console the client and to offer practical advice (on pet replacement, burial or disposal).

Take time and listen, if you are involved in something else, drop what you are doing and be attentive to the client.

Be aware that clients may ask or confide something to you that they would not tell the veterinarian.

Stay with the client if you sense that they would appreciate your presence.

Remember...

...expressions of caring and concern by the veterinarian and his staff to a bereaved client are particularly important because society tends to be unsympathetic toward pet loss. Some pet owners are even embarrassed at the depth of their grief over the death of their animals, or think that other people will consider their behavior bizarre. In such cases, clients may not willingly express their true feelings. In light of what can be an unsupportive atmosphere for the bereaved pet owner, sincere and uncritical support from the veterinarian is our final responsibility to the client and an important factor in continuing a professional relationship with that individual.

References

1. Horn, J.C. and Meer, J.: The pleasure of their company. *Psychology Today* 18:52-58. 1984.

2. Schoenberg, B.M. (ed.): Bereavement Counseling: *A Multidisciplinary Handbook*. Greenwood Press. Westport, Connecticut. 1980.

3. Kubler-Ross, E.: *On Death and Dying*. Macmillan, New York, 1969.

4. Kay, W.J., Nieberg, H.A., Kutscher, A.H., et al (eds.): *Pet Loss and Human Bereavement*. Iowa State University Press. Ames, Iowa. 1984.

5. Dunlop, R.S. *Helping the Bereaved*. Charles Press Publishers. Bowie, Maryland. 1978.

6. Nieberg, H.A. and Fischer, A.: *Pet Loss, a Thoughtful Guide for Adults and Children*. Harper and Row. New York. 1982.

The preceding article is reprinted with permission of

Norden News, a publication of Norden Laboratories, a SmithKline Beckman Company, and of its author, who thanks Mark Dana for assistance in its preparation.

Dr. Garcia is a director of a clinic in Tampa. Florida. He is a graduate of the University of Florida and of Auburn Veterinary School. He has served four terms on the Florida Board of Veterinary Medicine and is a contributing author for several veterinary journals. Dr. Garcia is a member of the Advisory Board of Purina Practice.

Latham Letter readers are invited to communicate with Dr. Garcia at the Veterinary Medical Clinic, Inc. 4241 Henderson Blvd., Tampa, FL 33629.

The Latham Letter, Vol. VIII, No. 1., Winter 1986/87, pp. 10-13.

[1] Someone has estimated that our children witness 10,000 television deaths by the eighth grade. Perhaps the casual depiction of death on the screen contributes to poor understanding and acceptance of it in real life.

Grief Counseling for Euthanasia

Cecelia J. Soares, D.V.M., M.S.

In my work with grieving clients, both as a veterinarian and as a therapist, I have noted that the majority of people seeking my support in problematic grieving have been dealing with the issue of euthanasia. Some have felt the need for support in the decision-making process prior to euthanizing their pet, others have come for counseling because of unexpectedly strong reaction after the fact. Many clients in the latter group have had unrealistic expectations about the mourning period following pet loss; for example, they have been concerned that a week following the animal's death they were still grieving, often acutely. For these reasons, and because euthanasia represent a special problem in pet loss, I have decided to focus this paper on practical aspects of counseling for that issue, emphasizing my clinical experience.

TIMING

The major question facing clients anticipating euthanasia is how they will know when to have their animal euthanized.

Much of this particular decision-making process revolves around the reason euthanasia is being considered. In one study of patient contacts for euthanasia involving 23 veterinarians, McCulloch and Bustad (1983) reported that 3% of those contacts involved discussions of or the performance of euthanasia. Of those performed, 76% were done for clinical reasons, 9% for economic reasons, 8% for family reasons (death of family member, moving, etc.), and 4% for behavior problems[1]. It has been my experience that euthanasia decisions for clinical and behavioral reasons are the most problematic for clients in terms of timing, unless there is severe traumatic injury requiring an immediate decision.

Case #1: Mr. P, a middle-aged gentleman living with his very elderly and somewhat senile widowed mother, sought pre-euthanasia counseling services. His 10-year-old dog, "Ted," had been diagnosed as having prostatic cancer and had noticeably deteriorated in the past few weeks, though he was still eating and responsive in other ways. Mr. P's principal concern was whether Ted was suffering, he also wished to ease Ted's loss for his elderly mother as well as for himself.

In this case, on considering all possible alternatives, Mr. P chose to take advantage of specialty referral to a veterinary oncologist. This choice permitted him 1) to know that he had done everything possible for Ted, and 2) to have time to prepare himself and his mother more completely for Ted's death. In the process of counseling, he realized that much of his distress in contemplating euthanasia arose from his need to make the greatest possible effort on Ted's behalf. In fact, Ted underwent a period of treatment and was not euthanized until there was clear evidence of untreatable metastasis.

This case illustrates three important points relevant to the timing of euthanasia for clinical reasons: 1) the principal concern of the vast majority of owners is whether the pet is in pain--in some cases there is a clear cut choice due to the animal's condition, but many other cases fall into a "gray area," especially if the pet has relatively "good" and "bad" days; 2) many clients need an opportunity to "say goodbye"; many owners, in order to reduce the guilt which seems to be an almost inevitable accompaniment of euthanasia decisions, need to do as much as their personal resources will allow toward preserving their pet's life, even if a visit to a specialist, for example, simply confirms a grim prognosis. Other factors entering into the decision may be, for example, incontinence, deterioration in temperament, and possibilities for managing the animal physically.

Case #2: Mrs. M, a middle-aged woman with disabling intervertebral disc disease, presented for treatment a 15-year-old female Collie with severe bilateral decubitus over the greater trochanters. The dog was completely paralyzed in the hindquarters due to a degenerative disease of the spinal cord. She was also incontinent of urine and had urine scalding on the medial surfaces of both rear legs.

Although given a prognosis that no recovery was possible for the dog, and although urged by her husband and grown children to have the dog euthanized, she was unable to make that decision. Not until she had nursed the dog intensively over the next six weeks and was herself re-hospitalized with an exacerbation of her own spinal disease, was she able to authorize euthanasia. Her comment to her husband at the time was that she "had to do everything possible, no matter what" and that she "couldn't bear to bring her to the (veterinary) hospital and leave her there," but that it was acceptable if her husband did it because she was unable to.

Many people were involved in supporting this client's decision-making process--the professional and lay staff of the veterinary hospital concerned as well as the woman's family members.

One point that has emerged consistently for me in doing counseling for euthanasia is that the best time for the procedure is when the client is best prepared emotionally. Occasionally, of course, circumstances intervene, such as when there is a sudden trauma and an immediate decision is necessary. It is also crucial to clients that they have the most complete possible information. It is worthwhile to a veterinarian, even in the face of apparent client resistance to such information, to provide it. This may require a number of visits and/or telephone conversations.

TELLING FAMILY MEMBERS

It is most desirable that a client confer with family members concerning an impending euthanasia. Usually others, including children, are already aware of a pet's poor health. All family members need an opportunity to express their feelings about the pet's death.

Often a child's first experience with death occurs with the death of a pet. Children have openhearted attachments to pets, so when a loved animal dies, "... a whole world may open up to a child: a world of loss, grief, bereavement and confusion about how to handle this first experience with death"[2]. It is certainly crucial that children not be lied to or given fantasy stories about what happened; if this is done, the period of grieving is likely to be prolonged and the process complicated. I have seen in practice a number of adults who still remembered the childhood pain and confusion which resulted from well-meant adult prevarications about death of a pet (or, of course, of a human).

When an entire family can discuss the impending euthanasia of a pet, the possibility develops for mutual support in a profound way. I am aware of one family where the children (ages 9 and 14) actually encouraged and supported the adults in their difficult decision to euthanize an aged and debilitated dog.

GUILT

No matter how clear the choice in favor of euthanasia by objective standards, its aftermath is most often guilt and uncertainty. Pet owners need to realize that this is a natural reaction arising from a deeply ingrained principle with which we have all been raised--that killing is immoral[2]. It has been my experience in counseling owners, whether pre or post-euthanasia, that the more complete their intellectual and emotional preparation, the more able they are to handle their guilty feelings. In many cases, regrets surface concerning previous losses and their actions at those times.

Case #3: Mrs. E was referred to me for counseling by her veterinarian. He was concerned about her reaction following her decision to euthanize her late son's dog due to cancer of the bone. The dog had taken a slight fall, but due to the bone malignancy had sustained a fracture. In his view, euthanasia was a humane and necessary decision because of the extremely poor prognosis.

In meeting Mrs. E, I discovered that she felt particularly badly because she had not been with the dog when it was euthanized. She then revealed that she felt extremely guilty for also not being with her mother or daughter when they had died; the son had perished in an accident, so she had not been present at his death either. The painful feelings associated with those previous losses were very much a part of her reaction to the loss of the dog.

Many clients experience intensified guilt if they decide to euthanize their pet due to family reasons (moving, etc.) or because of behavioral problems. These clients tend to blame themselves for "taking the easy way out" in the former case, or for being responsible for the out-of-control behavior in the latter. Occasionally, such as the incontinent animal, a client will feel particularly guilty for making a euthanasia decision for reasons of inconvenience, even if it is extreme or interferes with care of other family members (e.g. very young children).

Case #4: Ms. T, a young woman who had recently relocated and was living with a relative, sought counseling after euthanizing her elderly dog. The euthanasia had been necessary due to pressure from the relative with whom she was living, as well as the slim possibility she could move to a rental unit which would allow dogs.

In the process of counseling she related that the death of her mother with whom she had a problematic relationship, had been mysterious and her grief about that death was in large part unresolved.

CLOSURE

One of the most important factors in mitigating grief after pet loss of any kind is the client's experience of closure at the time of death. Prior to euthanasia it is very important, if at all possible, for clients to have an opportunity to "say goodbye" to a pet. A last evening with the animal at home or a visit to the clinic may be sufficient; in other cases, clients and their families may need a longer time (see Case #2, above). If an animal has been in a veterinary hospital for some time, and in the veterinarian's estimation, euthanasia is imperative due to the animal's suffering and debilitation, then a final visit is likely to convince the owner and the family of the validity of that professional judgment.

There are two aspects to closure after the death of a pet. First is the immediate aspect of being present when the animal dies or seeing the remains of the animal after its death. Many owners need to see their pet's body to assure themselves that it is really dead [3]. As difficult as it may seem to veterinarians who are involved, it has been my experience that clients who are present at the euthanasia are less likely to have problematic grief reactions, including nightmares and unpleasant waking fantasies. Being present also defuses myths about the euthanasia process, although clients need information prior to the procedure to prepare them for the effects of the death process (e.g. agonal breathing).

The second aspect of closure has to do with putting the pet to rest. Distracted owners are often pressed into making hasty decisions about burial, cremation, or other disposal. It is most beneficial if owners are informed of their alternatives by veterinarians and/or counselors before their pet is euthanized. The client in Case #3, for example, experienced a significant decrease in her anxiety when she received the ashes from the crematorium and was able to dispose of them as she had planned.

Regardless of the method chosen for disposal of the animal's body, some type of family ceremony when a pet dies can provide a large measure of comfort for a grieving owner. This may include burying the pet, complete with headstone, or simply reciting a favorite poem or prayer with family or friends present. These ceremonies seem to be of additional importance for children.

COUNSELING BEREAVED PET OWNERS

All of the principles considered to apply to grief counseling for human loss apply equally to grief counseling for pet loss generally and pet loss due to euthanasia specifically.

1) First in importance, in my experience, is the acknowledgment of the real loss. They need to know that it is normal, in fact desirable, to grieve for the loss of any friend or companion--that it is a sign of the loving and caring relationship which existed between them. Most, if not all, clients have experienced limited understanding on the part of the people in their usual support network. Many of the clients I have seen have been asked after only one or two weeks, "You mean you're still upset over that? It was only an animal." One very comforting piece of information for clients in my experience is that the *average* duration of grieving after the loss of a pet is 10 months [4].

2) Most clients simply need to be listened to without judgment. If someone will listen in this way, the grieving owner can verbalize his or her painful feelings and, in this way, discharge them. Many of my clients have expressed their appreciation just for that aspect of counseling alone--the opportunity to be heard in a supportive atmosphere.

3) Clients need to understand the grieving process. Such books as Kubler-Ross' *On Death and Dying* or Nieburg and

Fischer's *Pet Loss* are wonderful resources for grieving pet owners. For children, I particularly like Viorst's *The Tenth Good Thing About Barney*, White's *Charlotte's Web*, and Jackson's *Telling a Child About Death* (an adult guide).

4) My experience has been that many owners dealing with pet loss wish to address the issue of pet replacement, particularly if there are children in the home. It is important for these clients to understand that the decision of whether or when to replace a deceased pet is very individual to that particular owner or family. Responses range from never wanting another pet to immediate replacement. If someone in the family is having difficulty accepting the death of an animal, in particular a child, obtaining another pet too early may seem disloyal to that person and may complicate the grief.

5) Appropriate referral needs to be made in selected circumstances. Reaction to the loss of a pet by an individual may fall anywhere along a range of possibilities: it may, for instance, be minimal, may bring up other suppressed losses in that person's life, or may even precipitate a severe depression with suicidal actions [5]. In one study of 52 bereaved pet owners, 10% reported intense, problematic grieving--particularly those with no children at home nor other companion animals [6].

In summary, it is important to establish an environment in which bereaved pet owners have the opportunity to grieve. This may take the shape of a pet-loss support group, individual counseling sessions, a concerned and compassionate veterinary staff, and/or loving and supportive family members and friends. The presence of such an environment in the life of the grieving person reduces distress and minimizes the possibility of prolonged emotional pain or psychosocial dysfunction.

References

1 McCulloch, M.J., & Bustad, L.K. Incidence of euthanasia alternatives in veterinary practice. In A.H. Katcher & A.M. Beck (Eds.). *New Perspectives on Our Lives with Companion Animals*.

Philadelphia: University of Pennsylvania Press, 1983.

2 Nieberg, H.A., & Fischer, A. *Pet Loss: A Thoughtful Guide for Adults and Children.* New York: Harper and Row, 1982.

3 Harris, J.M. Understanding animal death: bereavement, grief, and euthanasia. In R.K. Anderson, B.L. Hart, & L.M. Hart (Eds.). *The Pet Connection.* Minneapolis, Minn.: University of Minnesota, 1984.

4 Katcher, A.H. & Rosenberg, M.A. Euthanasia and the management of the client's grief. *Compendium on Continuing Education for the Small Animal Practitioner,* 1979, 1:887-890.

5 Keddie, K.M.G. Pathological mourning after the death of a domestic pet. *British Journal of Psychiatry,* 1977, 131:21-25.

6 Stewart, M. Loss of a pet--loss of a person: a comparative study of bereavement. In A.H. Katcher & A.M. Beck (Eds.). *New Perspectives on Our Lives with Companion Animals.* Philadelphia: University of Pennsylvania Press, 1983.

Cecelia J. Soares is a practicing veterinarian in Contra Costa County, California, has an advanced degree in clinical counseling and a special certificate in grief counseling. She teaches workshops for veterinarians, involving such topics as "interpersonal relations" and "stress reduction." Dr. Soares is a member of the Human/Companion Animal Bond Committee of the California Veterinary Medical Association.

Cecelia Soares: The Latham Letter, Vol VIII, No. 1, Winter 1986/87, pp. 1-4.

CHILDREN

Therapy Dog in the Classroom

What the Students Say

Dee Press, M.A.

Dee Press, M.A. is a teacher of emotionally disturbed children. She has produced a newsletter, Paw Power, for her class and in it the children described their feelings about the Therapy Dog, Echo, who assists Ms. Press in the classroom. We felt it would be of special interest to Latham Letter readers to view pet-facilitated therapy from a different perspective-- that of the children's. Ms. Press graciously granted us permission to reproduce the letters and drawings and provided an update which appears following the letters.

Echo is a highly trained Golden Retriever who has been coming to "work" in my classroom every day for over five years. She is a registered Therapy Dog through Therapy Dogs International.

My self-contained Special Day Class here at Camarillo State Hospital consists of 8 severely emotionally disturbed children who range in age from 9 - 13 years. They function in a second-third grade level, and they exhibit a wide range of extreme social and emotional problems. Their behavior may be intrusive, aggressive, defiant, destructive, withdrawn, or otherwise negative for a variety of reasons. They come to us when they have failed to respond to the best resources their communities could offer, including counseling, therapy, special teachers, and other professional intervention. Whatever their problems, their communities have determined they need treatment around the clock with specially trained staff to help them make a change in their lives for the better.

The Therapy Dog in our classroom greatly enhances warmth and therapeutic opportunity. Echo provides loving, non-judgmental interaction, sometimes called Pet-Facilitated-Therapy. She helps break through the isolation and loneliness common to emotionally disturbed children. The students here have had to leave their friends, family, and pets, and Echo's affectionate, consistent presence is obviously appreciated by them.

Echo is always nice to me. She is my friend, and I am her friend even when I have a bad morning. She is a special friend to me. When I am lonely, she helps me to feel better. She comes over to my desk when I am working and leans on me and makes happy noises, and then I pet her and we both feel happy. She teaches me to be kind and nice to people and animals, too. I feel happy when she wins trophies in obedience competition. She is the smartest dog I've ever known.

Mike

When I am sad, I like to sit on the bean-bag chair with Echo and hug her and pet her. She helps me to feel less lonely. My dog, Raggedy, died and Echo helps me to feel better about it. I like it when we help train her at recess and we see how to train dogs to do new things. I feel real happy when she wins trophies at dog shows. I help her to be a winner, and she helps me to be a winner.

Ricky

It is real special having Echo in my classroom every day. My desk is very close to Echo, and she helps me to feel comfortable and more at home. She helps me feel calm and less lonely. I like it when she carries that big soccer ball everywhere! It is so cute! She makes me smile. I like it when her ears go up when she is listening to Dee talk to her. I like to see her try and try until she gets two tennis balls in her mouth at the same time. I like it when she goes to the gym for PE with us, and she plays so nice and runs with all of us. I like to see her chase the ball and bring it back to us. I love Echo!

Brian

Echo is the best dog I have ever seen. She lets me hug and pet her. She helps me to calm down when I am upset. She gives me attention and I like that. I am real proud of her for winning trophies. She makes me happy every day. I watch Dee comb her so I learn how to take care of my dog when I get home.

Alex

When my dog "Baby" died, I was very upset for a long time. Now Echo is a very special friend to me. I like to play with her. I like it when she wins trophies at dog shows. She is the only dog I know that is trained so well. When I get older, I would like to have my own dog and train it to be in dog shows too!

Angie

I like having a dog in my classroom. It makes me feel happy when she gets so excited when she first sees me. I like it when she comes over to me and lets me hug her for a long time! I like the way she stays right with Dee and listens to the commands during the obedience exercises. She makes me want to listen to my teacher and do good, too!

Mike

Echo is a very nice dog. I love to play with her so I try to earn all my points so we can play at the end of the day. She helps me to be good. I learn not to tease. I learn not to give them candy and to give them the right kind of food. I learn how to teach dogs things. It's fun to see her chase the ball and frisbee. I like to see Dee and Echo do obedience training. She is a special friend to me.

Martille

I think Echo is so cute! I love to play with her. She always comes over to see me. I like to give her hugs. I'm very happy when she wins stuff at dog shows. She is nice to us at recess. I have a dog at home, and Echo helps me not to miss my pet so much because she is with me every day at school.

Tony

Echo is one of the smartest dogs I know. She does good tricks, and I am proud that she wins ribbons and trophies at the dog shows in obedience competition. She is cute and I love it when she greets us and makes that cute little noise with her voice. She makes me feel good. It's great to have a pet in the classroom.

Jeremy

Echo is a special friend to me. I like playing with her and giving her hugs. I learn not to feed dogs junk food, not to tease them, and to say, "Good Girl" when she does what I ask. Even though she is not a cat, she helps me forget that my cat, "Tigger," died.

Jason

I like Echo. I like to sit on the bean-bag chair and brush her fur. She likes me. She is pretty and she is nice to me. She is smart, and she does good in dog shows. I like to talk to her. She always listens to me.

Leah

I love to watch her do obedience exercises. She makes me believe in myself, and she makes me happy. I like it when she obeys Dee. I like it when I pat her and she wags her tail. It makes me laugh. I help her do "figure eights" by standing still with someone else so she can go around us. She helps me calm down, and I like to hear her "talk" and "read."

Sylvia

I was on vacation during July and August, 1990. While I was at home Echo died of a liver ailment in mid-July. She was 6 years old.

Having experienced her special contribution to so many in her role as Therapy Dog, I decided to get another Golden Retriever puppy right away while I still had a month and a half to start training and bonding before I returned to work in September.

During this time I communicated Echo's death first to the psychologist and interdisciplinary team that work with the students in my classroom, and then to the children through letter and later in person with the new therapy puppy: Echo (II) was 3 months old when I got her.

So -- even Echo's death was therapeutic in that the children could deal with the loss of their special friend with support of the therapists and teacher through the grieving process. Many had the opportunity to share (often) repressed feelings dating from loss of their pets, and some related feelings about loss of a friend or relative. They sent me cards and pictures and helped me with my grief, too. It was a valuable growth experience for us all, and I'm sure none of us will forget the influence of our first Therapy dog.

A few weeks ago, the new children in my class wanted me to do another Paw Power edition, but I have not yet found the time to formalize it. I'm including their latest contributions.

I knew our first therapy dog, (Big) Echo, for two years. She was in school with me every day. I thought she was the smartest dog in the world. She helped me calm down when I was "hyper." She was nice to pet when I was sad. I liked it when she ran after the ball.

When she died while Dee was on vacation I felt sad and I felt sorry for her. But I am glad she is happy again because she got a new dog and we call her Echo, too.

I think right now our new puppy is hyper and we are learning to train her. I learn that puppies act different than older dogs just like babies act different than I do.

We get to see our puppy get new teeth. We got to see Dee pull a tooth that was falling out. I think it's funny that she puts stuff in her mouth. Sometimes she runs real fast in circles in the playground and it makes me laugh. Soon we will teach her to run and catch the frisbee. She makes my days more happy.

Bruce

I like having a Therapy Dog in our class. I like it when she jumps after the ball. I loved our first Therapy Dog very much. When she died I was very sad because she was a special friend to me. She helped me learn to be kind to animals. She made me feel happy. I am very happy we have a new therapy puppy now. When my behavior is good, I get to help train her so my behavior gets better, too.

Otis

I didn't know the first dog. I'm sorry she died. I like having a therapy puppy in our class. She likes to play with me. I like the way she jumps around and chases her ball. I like to play with her at recess.

She is so soft, and she likes to be hugged. She makes me feel warm and calm when I hug her. She gives me puppy kisses.

I like our special time when I sit on the rug with her and brush her. Her fur is soft and warm.

I never got to have a pet before, so I am happy I have a special puppy that I can love in our classroom every day. She cheers me up and makes me want to do my best so I can play with her at "free time."

I like the way she is learning to obey rules. I learn to obey the rules, too, so I can earn time with her and go home with better behavior, too!

Tony

Our first therapy dog knew a lot of tricks. She was calm and gentle. She did everything the first time you asked her. She helped me know animals have feelings, too. She played 4-square with us at recess. She chased the ball for us but she was careful not to get in our way. When she died I felt real sad that she wouldn't be with us any more. Dee surprised us with a new therapy puppy and I felt excited about our new Echo. Our puppy is friendly and soft. We will train her to do good in dog shows, too.

Nahshan

I learned from our first therapy dog how smart and special dogs can be when they are trained with love and respect. She made me feel like I wanted to be good so she would be my special friend. Our new therapy puppy is a little hyper, but she is learning fast. We all help Dee to train her. We will teach her to "sing" with us in the Christmas show when our class sings "How Much is That Doggie in the Window?" She is learning how to do what we say. My behavior has to be calm, too, so that she will settle down, so we help each other to have good behavior.

Antwaren

Echo teaches me not to tease, to be calm around animals, and to be kind and respect animals and others. I learn by watching Dee how to train dogs so I can train my pet when I get home.

I felt sad and sorry for me and Dee when our special friend died. I hoped Dee would get another therapy puppy so we could have more fun in our class.

I don't feel so lonely with our new puppy in class. She is playful and we help to teach her not to jump and to sit before we pet her.

I like to see her get happy when she sees me. She makes me feel happy.

Jonathan

Dee Press, M.A. has been working with children at Camarillo State Hospital for 28 years. She is an Educational Therapist and Master Teacher and has supervised students from many colleges and universities. Echo was her first dog. Ms. Press stated: "My work with my therapy dogs has been the most professionally rewarding of my career in working with these special children who need so much to learn compassion and respect for all living things--and for themselves." Echo II is also registered as a Therapy Dog through the Delta Society's "Pet Partners" program.

The Latham Letter, Vol. XII, No. 1, Winter 1990/91, p. 1, pp. 14-16.

A Time of Innocence-
A Time of Violence

Marcia Kelly

Deep in the shadows of childhood, hidden among the nursery rhymes and stuffed animals, lurks a monster. Neither a storybook beast nor an imaginary under-the-bed denizen, this phantom is real and haunts millions of children. Its terrifying face may be as familiar as a mother's or father's or as unknown as a stranger's. It is the face of child abuse and trauma.

For many children, the carefree innocence of youth has, quite literally, taken a beating. In 1985, 2 million child abuse and neglect cases were reported nationally. Of these, 15,460 occurred in Minnesota, according to statistics from the Minnesota Department of Public Welfare.

As reporting requirements have developed over the years, the number of cases of suspected abuse has risen dramatically; the number of programs to help abused children, however, has not. Between 1981 and 1985, for example, the reported cases rose 54%, while resources rose 1.9 percent.

The result is a population of hurting children and limited help. "The crisis is, we've put the emphasis on the reporting," says Dr. Robert ten Bensel, professor of public health and nationally recognized expert on child abuse. "What are we doing to protect children and help families?"

Ask Dr. George Realmuto. In 1984, as the director of the University of Minnesota's inpatient child psychiatry unit, he noted that a large percentage of children he was seeing had been abused. He also saw that physicians, therapists, and social workers had no coordinated way to treat both the children and the situations from which they had come. Once the children left the hospital's care, they were sent home<197>often to the same people who had abused them initially.

Thinking that there had to be a better way, in 1985 Realmuto submitted a proposal for a child abuse center to the hospital's Clinical Program Development Fund, which provides start-up monies for innovative programs. Administrators were impressed: Realmuto's multidisciplinary approach, employing a psychiatrist, a psychologist, a social worker, and a psychiatric nurse was unique, and it positioned the clinic as a community resource. Convinced that such a clinic would fill a need and that it could become self-supporting, they granted the funds, thus launching the Comprehensive Clinic for Abused and Traumatized Children (CCATCh).

Besides having an innovative structure, the clinic also takes an atypical diagnostic approach. When children who have experienced personal disasters, a family member's death or their parents' divorce, for example, exhibit such "behavior problems" as depression, conduct disorders, hyperactivity, suicidal tendencies, they're referred to the clinic to be evaluated for post-traumatic stress disorder.

Often applied to Vietnam war veterans or concentration camp survivors, the term refers to a set of beliefs and behaviors typical of people who have experienced extreme and inhumane treatment. Adults, for example, might experience flashbacks, feel extremely detached and be estranged from people, be hyper-alert, suffer sleep disturbances, or have trouble concentrating or remembering.

Little has been written about the disorder in children but Realmuto has recognized that youngsters have a set of symptoms comparable to those suffered by adults. Depending on the child's developmental stage, they may include reenacting the abuse through play, having bad dreams and night terrors, believing they'll die young, and having cognitive malfunctions, such as time distortions.

The clinic is also unusual in its comprehensive methods of treatment. Parents are helped to identify and meet their own needs, and social-service organizations are consulted to determine whether outside intervention is needed. The main emphasis of the therapy, however, is to empower the children by teaching them how to exert some control over their often-chaotic lives.

One hundred and twenty clients have passed through the clinic's doors since it opened in March, 1986. The 67 children who participated the first year were there to deal with a variety of issues: divorce (43 cases), being a victim of or witness to violence (39), sexual abuse (29), neglect (22), death (11), separation from parents due to placement outside the home (17).

About half of the clinic's referrals come from social workers in the seven-county metropolitan area. A quarter come from the courts, and another quarter from schools, private physicians, foster-parent organizations, the Red Cross, or from self-referral.

PORTRAITS OF PAIN

"The bulk of our kids are between four and six, and eight and 10 years old," says Kristie Sward, clinic coordinator. "The split is about half girls, half boys."

Not all of the children have been physically or sexually abused, however. Some have been traumatized emotionally. They may have been neglected or been victims of natural disasters. Or they may have witnessed their mother being battered, or seen their house burn down, or been part of a vicious custody battle, or had someone they loved die. Their wounds, though invisible, are as devastating as the blows from an angry fist. Whatever the source of the suffering, "We try to think about all of it as trauma," Sward says.

As part of their therapy, the children are asked to draw pictures of events they can tell a story about, or of various emotions. Their artwork tells of feelings that words can't always convey. "S-A-D" spell the crayon letters in the upper right-hand corner of one picture. Beneath them is drawn a pigtailed pixie with a river of tears coursing down her cheeks. Another self-portrait is nothing but a chaotic jumble of overlapping shapes, an almost unrecognizable abstract of a child's sense of self.

For many of these children, the pain reflected in the drawings is but the smallest hint of their problems. More obvious signs show up in their behavior, which often is what prompts parents or other referral sources to seek out the clinic.

HEALING THE HURT

The sad fact about abuse and trauma is that the emotional scars never fully go away. Unlike depression or phobias, "post-traumatic stress disorder is not curable," Realmuto says. "In fact, it's recyclable," always able to be triggered by general stress over the course of one's life."

As a results, "We don't think we cure kids," Realmuto says. "We teach management [of symptoms and] ways not to be helpless."

The process begins by understanding that a child's perception of a traumatic event is not the same as an adult's. To illustrate the point, Realmuto tells of two Indian boys, ages four and five, who were adopted by a Minnesota family and

later abused by the family's oldest son. When the truth finally came out and the little boys were asked to identify the worst part of the ordeal, their answer surprised everyone. It wasn't the sexual abuse; it was the older boy's threat that if they told, they'd be kicked out of the family, and the knowledge that if they didn't tell, the secret would keep them from ever feeling connected to the family.

Sward confirms the observation. "Kids' perceptions of what's awful are different from adults'. It's not the sexual act that's horrible; it's bearing the secret."

CCATCh helps lighten the load. Initial assessment interviews give children the chance to tell their awful secret—either directly or through drawing or by using anatomically correct dolls to explain what happened. One method that's particularly useful is the trauma interview, a structured sequence of activities and discussions with a therapist.

It begins with the child drawing a picture and telling a story about it. From that exercise, the therapist gets insights into how the child is handling the trauma situation: Does he deny what happened, reversing the violent outcome? Does he avoid all emotion in his drawing and merely report journalistically? Is he anxious just talking about it?

The next step is to help the child face feelings and memories. Talking about what happened, affirming the child's physical and emotional reactions, discussing the worst moment of the abuse or trauma, and acknowledging fears of future abuse all help to heal the terrible pain and isolation that come from keeping a bad secret. It's important to "give the kids enough time to talk about what worries them the most," Sward says. Often it's abandonment.

After the trauma interview, the CCATCh staff determines if additional hospital services are required. An abused child might need a gynecological exam, psychological testing, or help with developmental delays. The staff also makes recommendations to social services if they think foster care should be considered or parental rights terminated.

The next step is to begin treatment, which typically lasts six months but may go on for a year. Some children get individual therapy, but most join groups of their peers.

"It's most effective to work with groups," Sward says. "They can share with others who have had similar experiences and feelings. They can know they're not alone."

They can learn how not to be victims. The goal is to "teach safety and responsibility, to give them a sense of power," Sward says. "These kids haven't seen adults handle kids' problems responsibly," she adds. "They need to learn coping skills. They need to learn how to use adults as resources and make reasonable attachment without being dependent or ripped off."

That's not always an easy lesson. For one young teenager it's difficult indeed. The child of a prostitute and pimp, she was permanently removed from her home and placed in foster care after her father was jailed for murder. Though subsequently adopted, she was again placed in foster care because of conflicting values and expectations between her and a very conservative family. By the time she was adopted by a second family, the girl's fear of strangers and sense of loss were enormous. The pain manifested itself in acts of aggression—picking fights when the family would try to get close to her, for example, or threatening to run away.

To teach the girl how to control her fear, grief, and anger, Realmuto says, CCATCh staff worked with the adoptive parents, educating them about post-traumatic stress disorder and suggesting ways they could give the child a sense of power. First they explained that the girl's aggression wasn't a personal affront, but rather the only way she knew how to protect herself from the pain that always followed emotional closeness. They also explained that the child's desire to find her siblings was not a rejection of her new family, but instead was an attempt to minimize some of her emotional losses.

Realmuto recommended that the parents support the girl in her search and empower her by suggesting ways she could help herself. The parents could, for example, encourage her to write to the social worker who had recommended terminating parental rights or call a previous foster family.

Evaluating parents and teaching them how to handle both their own and their children's problems is an integral part of the CCATCh program. "What we're looking for is their own family-of-origin history, family history (that influences the child, such as the death of a parent), development history, and how they go about parenting," Sward says, adding that whether the parents come to the CCATCh clinic voluntarily or because of a court order "colors the whole scenario."

People who have abused their children include "regular middle-class types, prostitutes, multiple felons, utterly neglectful parents, people who see their own lot in life as being beaten down," Sward says. "They feel like complete, utter life failures and complete failures as human beings."

CCATCh therapists, through counseling and videotaping interactions between the parent and child, help the parents identify the problems that are causing them to lash out at their children. It may be identity issues, unresolved grief and loss, victim issues, or job problems. "We give them hope about self-improvement and access services for them," Sward says.

If the parents' problems are severe or insurmountable, "the kids may not have time to wait," she adds. Foster care or termination of parental rights may be necessary.

Measuring success in a program that teaches coping skills is difficult. There are, however, six-month follow-ups with CCATCh clients that include questions about whether the child's symptoms are better or worse and whether the parents are able to monitor their own problems and reactions. Responses so far have been encouraging.

Despite a lack of quantitative data, one thing becomes clear: A light has been cast into the shadows of childhood, and the monster has been summoned to leave.

The foregoing article is reprinted with permission from the magazine, Health Sciences. It was initially published in the Winter 1988 issue of that University of Minnesota-sponsored publication.

Its author is editor in the office of Health Sciences Administration. Marcia Kelly holds a Master's degree in journalism. She is widely appreciated for her ability effectively to communicate important concepts in the fields both of child abuse and pet facilitated therapy.

The Latham Letter, Vol. IX, No. 4, Fall 1988, pp. 1, 15-18.

Reaction of Infants and Toddlers to Live and Toy Animals

Aline H. Kidd, Ph.D.
and Robert M. Kidd, M.A., M.Div.

Six-month-old Kori is propped on the floor, showing off. A lively cockapoo rushes up to him, sniffs and licks and then sits a few paces away, wagging tail, tummy, and head at him. Kori flops to his tummy and tries to crawl to the dog, clutching and crowing with delight. Such behavior evidences his interest in forming emotional attachments with other people, animals, or objects.

Our earlier study showed that most 3-year-olds were already attached to their pets. We also found that no one had yet actually observed the proximity-seeking and contact-promoting behaviors of infants with pets, so we began the present observational research of infant reactions to both live pets and toy animals.

Because of local environmental living realities, our 250 infant subjects came from intact suburban families, 134 of which had dogs, 84 had cats, and 32 had both dogs and cats. Each 6 through 30-month-old infant was visited at home with the family pet and one or both parents present. Each was observed reacting with, in random order, a mechanized toy dog that moved and barked realistically, with a cuddly toy cat that only purred and meowed realistically when hugged or petted, and with the familiar family pet. Our checklist of attachment behaviors included the amount of time spent with the family pet and with each of the toy animals, the amount of smiling, laughing, verbalizing, clutching, holding, playing and following or attempting to follow the pets and toys. We also noted any rejections such as pushing away, ignoring, or moving away from the pet or toy by the child.

The babies' attitudes toward the live pets and the toys differed more and more as the infants aged from 6 to 30 months. The 6-month-olds spent as much time and effort with the toy animals as with the live pets, many acting like the baby already described. The older infants, however, watched and responded more to the live pets, and the amount of time spent with the live pets increased steadily from 12 to 30 months. The amount of time spent with the toy animals did not change at all. Fifteen parents commented especially that their babies showed an increased interest in the family pet when they began to toddle and move more competently on their own at about fourteen months. Our normal, curious-about-the-world-about-them infant subjects certainly preferred interacting more with the animal pets who were their same size and on their same space level and with whom they could exchange friendly growls and meows while eye-to-eye.

There were a number of differences between boys and girls. During the first year, the boys laughed and talked more to and about their pets. They also touched and followed more, and permitted more contact with pets. The boy's and girls' reaction were equal at 18 months, but the 24 and 30-month-old girls showed more attachment behaviors. Of course, boys are more active during the first year than girls and so would react more to pets.

Because parents usually encourage girls to talk more than boys, the girls talked more from 24 months on to their pets and people. Interestingly, both sexes under 24-months usually referred to their pets as "doggy" or "kitty" and the 24 and 30-month-old toddlers usually called to the pets by their given names: "my Sparky, my Garfield, my O.J." All of the subjects referred to the toy animals as "doggy" or "kitty"; none ever gave them a name or asked if they had names!

Up to 24 months, the girls more often tended to push their pet away. Because fathers usually roughhouse with sons but not with daughters, and mothers rarely roughhouse with either, boys are more accustomed to strong physical stimulation and so are more likely to tolerate a pet's licking, pushing, and knocking them down than are girls who rarely experience roughhousing. After 24 months, however, both sexes achieve some ability to manage their pets, adjust to pet behaviors, and move capably enough to fend off pet attentions. So, most 2-year-olds are more companionable with their pets and rarely push them away.

We considered texture, sound, novelty, and movement as possible reasons why infants preferred live pets to to animals. Touch apparently had little effect on infant attachment behaviors. All possible coat textures were noted among the live pets to whom the babies were attached, and the infants did not react differently to the stiff curly coat of the mechanical dog or to the soft silky fur of the toy cat.

Sound, too, apparently played a minor role in child/pet attachment. Although the mechanical animals barked, purred, and meowed appropriately, neither the mechanical or real animal noises seemed to hold the babies' interest or attention significantly.

Similarly, the infants preferred the familiar family pet to the novelty of the toy animals, probably because they felt more secure with parent and family pet present when a stranger and strange toys were also present.

Probably the chief reason for preference of live pet over toy animal was responsive movement. The toy dog moved in set repetitive sequences, the toy cat merely made sounds when acted upon, but the live pets moved in response to the infants' behaviors toward them and, through their own movements, interacted with the infants. We know that in parent/child relationships, attachment is based on interactions between two individuals. The same appears to be true of child/pet relationships.

Our study makes clear that young children show much more attachment to family pets than to toy animals, and distinguish between pet types, preferring dogs to cats, as early as one year of age. This is, however, the first study of infant reactions to pets and we need to know a great deal more before we can fully understand infant/pet attachment behaviors. The influence of parental attitudes and the influence of pet breed and sex on infant behaviors are unknown, as are the effects of ethnic and cultural group differences on infant/pet interactions. Future studies will probably answer these and other questions raised by our study.

Pets and the Socialization of Children

Michael Robin

Robert ten Bensel M.D.

ABSTRACT

Despite the widespread ownership of pet animals in American families, there is very little analysis of the role of pets in child development. This paper will examine the influence of pet animals on child development; the impact of pet loss and bereavement on children; the problem of child cruelty to animals and its relationship to child abuse; and the role of pets in both normal and disturbed families. The authors will also review their own research study of adult prisoners and juveniles in institutions in regard to their experiences with pet animals.

INTRODUCTION

Given the large numbers of children who have had pets, it is striking how little attention has been paid to the role pets play in the emotional and developmental lives of children. In addition to the mythological, symbolic and utilitarian aspects of the animal/human relationship, recent research has focused on the developmental aspects of this relationship. While there is literature on the role of animals in myths, fairytales, dreams and nightmares, very little has been written on companion animals and children. This paper will focus on what is known about the normal developmental interactions between animals and children and the implications of this knowledge to the everyday lives of children. In addition to a review of the literature on companion animals and children, we will also report on our surveys of juveniles and adults in correctional institutions and their experiences with pet animals (Robin, ten Bensel, Quigley and Anderson, 1983, 1984; ten Bensel, Ward, Kruttschnitt, Quigley and Anderson, 1984).

COMPANION ANIMALS AND CHILDREN

Companion animals are a vital part of the healthy emotional development of children. As children develop, animals play different roles for the child at each stage of development. The period of childhood encompasses a number of developmental tasks - the acquisition of basic trust and self-esteem, a sense of responsibility and competence, feelings of empathy toward others and the achievement of autonomy - that can be facilitated for the child by a companion animal. The constancy of animal's companionship can help children move along the developmental continuum and may even have an inhibiting effect toward mental disturbances (Levinson, 1970).

In what ways can a pet meet the mental health needs of a child? In the first instance, a pet is an active and energetic playmate, which facilitates the release of a child's pent-up energy and tension (Feldmann 1977). In general, a child who is physically active is less likely to be tense than one who is not. The security of the companion animal may encourage exploratory behavior, particularly for fearful children in unfamiliar situations. It may also serve as a bridge or facilitator towards relationships with other children. And for those living in situations without other children, a pet may be a substitute for human companionship. As one child said, pets are important especially for kids without brothers and sisters. They can get close to this animal and they both can grow up to love one another (Robin, ten Bensel, Quigley and Anderson, 1983).

Caring responsibly for a pet will help a child experience the pleasures of responsible pet ownership. Levinson (1972) suggests that responsibility for pet care should be introduced gradually and that parents should recognize there will be periods when even for a conscientious child the care of a pet will be too much. Adolescents living in normal family environments more often shared the responsibility of pet care with other family members which became a source of mutual enjoyment (Robin, ten Bensel, Quigley and Anderson, 1983). The successful care of a valued pet will promote a sense of importance and being needed. By observing the pet's biological functions, children will learn about sexuality and elimination (Levinson, 1972; Schowalter, 1983).

In laboratory experiments, it was found that people of all ages, including children, use animals to feel safe and create a sense of intimacy. As Beck and Katcher (1983) have noted, pairing an animal with a strange human being apparently acts to make that person... or the situation surrounding that person, less threatening. For example, in an experiment where children were brought into a room with an interviewer alone or with an interviewer with a dog, the children were found to be more relaxed as measured by blood pressure rates when entering a room with the interviewer and an animal (Beck and Katcher, 1983). In another study in England, Messant (1983) found people in public parks were considered more approachable for conversation when accompanied by a pet. In general, the presence of companion animals seems to have a relaxing and calming effect on people. When people talk to other people there is a tendency for blood pressure to rise; however, when people talk to or observe animals there is a tendency for blood pressure to lower.

PETS AS TRANSITIONAL OBJECTS

It is widely accepted that the key factor in the relationship between children and companion animals is the unconditional love and acceptance of the animal for the child, who accepts the child as is and does not offer feedback or criticism (Levinson, 1969, 1972; Beck and Katcher, 1983). As Siegel (1962) has written, The animal does not judge but offers a feeling of intense loyalty ... It is not frightening or demanding, nor does it expose its master to the ugly strain of constant criticism. It provides its owner with the chance to feel important. The simple, uncomplicated affection of an animal for its master was also noted by Freud in a letter to Marie Bonaparte, It really explains why we can love an animal like Topsy (or Jo-Fi) with such an extraordinary intensity: affection without ambivalence ... that feelings of an intimate affinity, of an undisputed solidarity. Often when stroking Jo-Fi, I have caught myself humming a melody which, unmusical as I am, I can't help recognizing as the aria from Don Giovanni: A bond of friendship unites us both. (Freud, 1876).

PETS AS PARENTS

Beck and Katcher (1983) have suggested that as children get older, the pet acquires many of the characteristics of the ideal mother. The pet is unconditional, devoted, attentive, loyal and non-verbal - all elements of the primary symbiotic relationship with the mother. From a developmental point of view, a major task of childhood is the movement away from the primary symbiotic relationship with the mother and the establishment of a separate and distinct identity (Erickson, 1980). This process of separation and individuation creates feelings of separation anxiety that occur throughout the life process, particularly at stressful times of loss or during new experiences (Perin, 1983). One could regard the entire life cycle as constituting a more or less successful process of distancing from and introjection of the lost symbiotic mother, and eternal longing for the actual or fantasied ideal state of self (Mahler, 1972).

Pets function, particularly for adolescents, as transitional objects, much like the blanket or teddy bear does for infants. As transitional objects, pets help children feel safe without the presence of parents. Pets are more socially acceptable as transitional objects for older children than are inanimate objects. Adolescence brings with it a changing relationship to pets, in large part due to this emergence of pets as transitional objects. At this period pets can be a confidant, an object of love, a protector, a social facilitator or a status symbol (Fogle, 1983). Moreover, the bond between children and pets is enhanced by its animate quality. The crucial attachment behaviors of proximity and caring between children and pets forms an alive reciprocating alliance (Bowlby, 1969). The relationship is simpler and less complicated than are human relationships.

Like other transitional objects, most of the shared behaviors between animals and children are tactile and/or kinetic rather than verbal. Levinson (1969) has stated that pets may satisfy the child's need for physical contact and touch without the fear of entanglements that accompany contact with human beings. Children have a great need for empathetic listening and association with others. It is the non-interventiveness and empathy that makes animals such good companions. Pets are often perceived by children as attentive and empathetic listeners. As one child wrote, My dog is very special to me. We have had it for seven years now. When I was little I used to go to her and pet her when I was depressed and crying. She seemed to understand. You could tell by the look in her eyes. (Robin, ten Bensel, Quigley and Anderson, 1983).

PETS AS CHILDREN

Along with the parental role, pets simultaneously or alternately function as children for the pet owner (Beck and Katcher, 1983). This idea was expressed by the prophet Nathan during antiquity (2 Sam, 12:3): The poor man had nothing save one little ewe lamb, which he bought and nourished up; and it grew up together with him, and with his children; it did eat of his own morsel, and drank of his own cup, and lay in his bosom, and was to him as a daughter. Midgley (1984) notes in her discussion of this passage that the lamb was not a substitute for the poor man's children as he had children. His love for the lamb was nonetheless the kind of love suited to a child. The lamb was a live creature needing love, and was able to respond to parental cherishing. The helplessness of the animal drew out for the man nurturing and human caring.

Fogle (1983) notes that studies in New York State show that pets can elicit maternal behaviors in children as young as three years old. In fact, according to Beck and Katcher (1983), much of the usual activity of children and pet animals resembles a parent/child relationship with the animal representing the child as an infant. Children unconsciously view their pets as an extension of themselves and treat their pets as they want to be treated themselves. This process is what Desmond Morris has called infantile parentalism, suggesting this is one way children cope with the loss of their childhood (Morris, 1967). Schowalter (1983) for example, discussed the case of a five-year-old insecure boy referred for psychiatric care due to his habit of petting his goldfish. For this boy, petting the fish helped him feel both caring and cared for. Gradually he was able to transfer his affection toward a dog. With increased parental nurturance, he became more confident and outgoing.

Sherick (1981) also presented a case of a nine-year-old girl whose pets became symbolic substitutes for her ideal self. The sick pets that she cared for and nursed back to health represented the cared-for, protected and loved child that she longed to be. The girl's mother was a vain woman concerned with appearances who turned most of her maternal instincts toward the family pet rather than her daughter. The girl's behavior toward her pet was an unconscious effort to model good enough mothering to her mother. Searles (1960) points out that many children grow up with parents unable to nurture them, because of their own disturbance, but who can show affection to an animal. The child then grows up thinking if he or she were an animal then they might receive parental love. Kupferman (1977) presented a case of a seven-year-old boy whose ego development was so faulty that he took on the identity of a cat and meowed to his psychiatrist.

PETS AND FAMILIES

The role of a pet in a family will be dependent upon the family's structure, its emotional undercurrents, the emotional and physical strengths and weaknesses of each of its members, and the family's social climate (Levinson, 1969). When a pet is acquired by a family a variety of changes frequently occurs in family relationships and dynamics. Cain (1983) found in her study of pets in family systems that families reported both positive and negative changes after acquiring a pet. Some families reported increased closeness expressed around the care of a pet, more time spent together playing with a pet, more happiness of family members, and less arguing. However, other families reported more arguing and problems over the rules and care of the pet and less time spent with other family members; for example, children spent less time with their parents and husbands spent less time with their wives (Cain, 1983).

Pets become, according to the theory of Murray Bowen, part of the undifferentiated ego mass of the family and form part of the emotional structure of that family (Bowen, 1965). Many people indeed consider their pet as a member of the family. In Cain's survey of 60 families, 87 percent considered their pet as a member of the family (Cain, 1983). Ruby has also noted that most families include their pets in their family photographs (Ruby, 1983). Family members not only interact with their pets in their own characteristic manner, but they also interact with each other in relationship to the pet. In some families, pets become the major focus of attention and assume a position even more important than family members (Levinson, 1969).

182   Pets and the Socialization of Children

As Levinson has cautioned, pets may be involved in family pathology (Levinson, 1969). For example, one young woman committed suicide after being ordered by her parents to kill her pet dog for punishment for spending the night with a man. The woman used the same gun on herself that she used to kill her dog (Levinson, 1969). In another case, Rynearson (1978) discussed a severely disturbed adult woman who as a child had a profound fear of her parents and siblings. She turned to her cat as a confidant with whom she shared her troubles. One day her younger sister was scratched by the cat and the woman watched her enraged mother kill the cat with a shovel and then her mother turned to her and said, "Never forget that you are the one who really killed her, because you weren't watching her closely - it's all your fault."

Children can involve their animals psychodynamically in their use of such defense mechanisms as displacement, projection, splitting and identification (Schowalter, 1983). There are times when a child living in a disturbed family will become overly attached to a pet to the detriment of human relationships. Such children have a basic distrust of people which becomes overgeneralized. This basic distrust of human attachments contributes to the intense displacement of attachment to a pet who is consistently receptive as a source of love and caring. In anxiously attaching to an animal, a child can gratify part of the self without risking interpersonal involvement. Disturbed children with limited ego strength will turn to their pets for warmth and caring to meet their regressed, insatiable need for closeness and love (Rynearson, 1978; Levinson, 1972).

In a study of 269 disturbed children institutionalized for delinquency problems, 47 percent said pets were important for children growing up because they provided someone for them to love. For the control group of students in regular public schools, a pet was important to them because it taught responsibility. For many abused and disturbed children, a pet becomes their sole love object and a substitute for family love. As one boy said of his pet, "My kitty was the joy of my life. It never hurt me or made me upset like my parents. She always came to me when she wanted affection." Another boy wrote, "My favorite pet was my dog Bell. I loved her very much. I took care of her all the time and never mistreated her. Sometimes she was the only person I could talk to." Overall, abused and disturbed children in this study were more likely to talk to their pets about their problems. Pets became their sole source of solace at times of stress, loneliness or boredom (Robin, ten Bensel, Quigley and Anderson, 1983).

PET LOSS

For many children, the loss or death of a companion animal is the first experience with death and bereavement. In fact, it is often stated that one of the most important aspects of pet ownership for children is that it provides the child with experiences of dealing with the reality of illness and death which will prepare them for these experiences later in life (Fox, 1983). By fully experiencing the grief of losing a pet, the child learns that death is a natural part of the life process, is painful, but is tolerable and does not last forever. A child can learn that death is permanent and that dead animals will not come back

184 Pets and the Socialization of Children

to haunt them. The children can also be taught that guilt feelings following the death of a loved object are common and can be overcome (Levinson, 1972).

There is a tendency, however, to minimize a child's grief over a lost pet. In the vast literature on children and bereavement there are few references to bereavement from pet loss (Nieburg, 1982). The death of a pet has been considered an "emotional dress rehearsal" and preparation for greater losses yet to come (Levinson, 1967). However, there are strong indicators that the loss of a pet is more than a "rehearsal," and it is a profound experience in itself for many children.

In a study of 507 adolescents in Minnesota, over one-half had lost their "special" pet and only two youths reported feeling indifferent to the loss (Robin, ten Bensel, Quigley and Anderson, 1983). Most of the youths whose pets had died had deep feelings of regret and sadness such as those who wrote, "My sorrows are very deep for my special pet, but I know she is in some place where she is treated very well. And I know she is thinking of me because I always think of her." And, "I was sad that he had to be put to sleep but I was glad that he didn't die painfully." Stewart (1983) also surveyed 135 schoolchildren in central Scotland on their experiences and feelings toward pet loss. She asked the children to write about their pets and how they felt if their pet had died. She found that 44 percent had pets that died and two-thirds of these children expressed profound grief at their loss, such as the child who said, "I didn't believe it, I didn't know where I was." In most cases, the children got over the loss, usually with parental support. But in all the bereavements that seemed unresolved the parents were unwilling to have another animal.

How a child reacts to the loss of a pet depends largely on his or her age and emotional development, the length of time the child had the pet, quality of the relationship, the circumstances surrounding the loss of the pet, and the quality of support available to the child. Pre-school children are less likely to view the pet loss as irrevocable. According to Nieberg and Fischer (1982), children under five years usually experience the pet loss as a temporary absence, and from five to nine years or so, pet loss is not seen as inevitable and is believed possible to avoid. Stewart (1983) found that school-aged children often expressed profound grief for a short time, and then seemed to quickly adapt to normal, especially if a new animal was introduced. Most young children miss their deceased animals, but more as a playmate than as an object that satisfies basic emotional needs.

It is usually adolescents who have the most profound experiences with pet loss. From early adolescence on, children begin to develop an adult perception that death is final, permanent and inevitable (Nieburg and Fischer, 1982). Adolescents tend to take longer to get over their grief, in part because their relationships with pets tends to be more intense at this age (Stewart, 1983; Nieberg and Fischer, 1982). How a young adolescent will react to pet loss will depend on the circumstances surrounding the death of a pet. A pet may be lost in a variety of ways such as old age or illness, being run over, theft, given away or traumatic death. Unfortunately, there are very few empirically based epidemiological studies on the nature of pet loss. In Minnesota it was found that abused and disturbed youths suffered more pet loss, had their pets for shorter times, and were most likely to have had their pet killed accidentally or purposely more than any other factor (Robin, ten Bensel, Quigley and Anderson, 1983, 1984). Most of those children whose pets were traumatically killed were saddened by the loss of their pet, and, in a few cases, were angry and revengeful toward the person who killed their pet. For example, one child wrote, "He was 11 years old and my mother had my little brother and Duke started being grouchy and nipping at people. So my brother-in-law shot him. It really hurt bad, like one of my brothers died. It was really hard to accept" (Robin, ten Bensel, Quigley and Anderson, 1983). Another child wrote, "My sister was taking it for a walk and this man drove over it, then backed over it and then drove over it again. I was hurt very bad. I hated that man. I cried for two days straight" (Robin, ten Bensel, Quigley and Anderson, 1983). Not only did abused and disturbed youths experience more traumatic pet loss than did the controls, they were also less likely to have someone to talk to about their grief. Only 56 percent of those youths whose pets died traumatic death had someone to talk to about their grief, as compared to 79 percent of the control group who had support after traumatic pet loss.

Most mental health practitioners indicate that the forms of bereavement from pet loss are similar to those of human loss (Levinson, 1967). Some children might be surprised and embarrassed by the intensity of their grief and feel the need to conceal their grief from the outside world. Parents should be sensitive to the child's grief and not minimize or ridicule its impact. Some young children tend to view the death of a pet as punishment from their misdeeds. If so, children should be assured that they were not to blame for their pet's death. Given that our society has no public rituals for the death of pets, families may enact funerals to acknowledge the importance of the pet to the family (Levinson, 1967; Nieberg and Fischer, 1982). Children should also be offered a replacement pet; however, there is disagreement if the replacement should be deferred for a time (Levinson, 1981; Nieberg and Fischer, 1982) or take place immediately (Stewart, 1983).

CHILDHOOD CRUELTY TO ANIMALS

Interest in childhood cruelty to animals grew out of the notion that cruelty to animals has a disabling effect on human character and leads to cruelty among people (ten Bensel, 1984). This idea was articulated by Saint Thomas Aquinas (1225 - 1274) who said: "Holy scriptures seem to forbid us to be cruel to brute animals ... that is either ... through being cruel to animals one becomes cruel to human beings or because injury to an animal leads to the temporal hurt of man" (Thomas, 1983). Likewise the philosopher Montaigne (1533 - 1592) wrote that "men of bloodthirsty nature where animals are concerned display a natural propensity toward cruelty" (Montaigne, 1953).

Until the seventeenth and eighteenth centuries, there was relatively little awareness that animals suffered and needed protection because of this suffering. This new sensibility was linked to the growth of towns and industry which left animals increasingly marginal to the production process. Gradually society allowed animals to enter the house as pets, which created the foundation for the view that some animals at least were worthy of moral consideration (Thomas, 1983). The English artist William Hogarth (1697 - 1764) was the first artist to both condemn animal cruelty and theorize on its human consequences. His Four Stages of Cruelty (1751) was produced as a means of focusing attention on the high incidence of crime and violence in his day. The four drawings trace the evolution of cruelty to animals as a child, to the beating of a disabled horse as a young man, to the killing of a woman, and finally to the death of the protagonist himself. As Hogarth declared in 1738, "I am a professional enemy to persecution of all kinds, whether against man or beast" (Lindsay, 19779).

The link between animal abuse and human violence has been made more recently by Margaret Mead (1964) when she suggested that childhood cruelty to animals may be a precursor to anti-social violence as an adult. Hellman and Blackman (1966) postulated that childhood cruelty to animals, when combined with enuresis and firesetting, were indeed effective predictors of later violent and criminal behaviors in adulthood. They found that of 31 prisoners charged with aggressive crimes against people, three fourths (N = 23) had a history of all or part of the triad. The authors argued that the aggressive behaviors of their subjects were a hostile reaction to parental abuse or neglect. Tapia (1971) found additional links between animal abuse, child abuse, and anti-social behavior. Of 18 young boys who were identified with histories of cruelty to animals, one-third had also set fires, and parental abuse was the most common etiological factor. Felthous (1980), in another study, found that Hellman and Blackman's behavioral triad did have predictive value for later criminal behavior. He found extreme physical brutality from parents common, but he felt that parental deprivation rather than parental aggressiveness may be more specifically related to animal cruelty.

Kellert and Felthous (1983) also found in their study of 152 criminals and non-criminals in Kansas and Connecticut

188 Pets and the Socialization of Children

an inordinately high frequency of childhood animal cruelties among the most violent criminals. They reported that 25 percent of the most violent criminals had five or more specific incidents of cruelty to animals, compared to less than six percent of moderate and non-aggressive criminals, and no occurrence among non-criminals. Moreover, the family backgrounds of the aggressive criminals were especially violent. Three-fourths of all aggressive criminals reported excessive and repeated abuse as children, compared to only 31 percent for non-aggressive criminals and 10 percent among non-criminals. Interestingly, 75 percent of non-criminals who experienced parental abuse also reported incidents of animal cruelty.

These studies identified extreme parental cruelty as the most common background element among those who abuse animals. As Erich Fromm has noted in his study, The Anatomy of Human Destructiveness (1972), persons who are sadistic tend themselves to be victims of terroristic punishment. By this is meant punishment that is not limited in intensity, is not related to any specific misbehavior, is arbitrary and is fed by the punisher's own sadism. Thus, the sadistic animal abuser was, himself, a victim of extreme physical abuse.

While most children are usually sensitive to the misuse of pets, for some abused or disturbed children, pets represent someone they can gain some power and control over. As Schowalter (1983) has said, "No matter how put upon or demeaned one feels, it is still often possible to kick the dog." Cruelty to animals thus represents a displacement of aggression from humans to animals. Rollo May (1972) suggests that when a child is not loved adequately by a mother or father, there develops a "penchant for revenge on the world, a need to destroy the world for others inasmuch as it was not good for him." Severely abused children, lacking in the ability to empathize with the sufferings of animals, take out their frustrations and hostility on animals with little sense of remorse. Their abuse of animals is an effort to compensate for feelings of powerlessness and inferiority.

A weakness of the previous studies of childhood cruelty to animals is that they did not consider the patterns of pet ownership among their subjects. These studies did not distinguish if the abused animal was the child's own animal or if the child had ever had a companion animal and what the nature of that relationship might have been. Other than a passing comment by Brittain (1970) in his study of the sadistic murderer, little mention is made of the child and his relationship to animals prior to the incident of cruelty. Brittain wrote, "There is sometimes a history of extreme cruelty to animals. Paradoxically they can also be very fond of animals. Such cruelty is particularly significant when it relates to cats, dogs, birds and farm animals, though it can also be directed toward lower forms of animal life, and the only animal which seems to be safe is the one belonging to the sadist himself." It is with these ideas in mind that we studied adult prison populations along with abused adolescents institutionalized for delinquency and emotional disturbances to determine their patterns of pet ownership and their feelings toward their pets.

In our study of 81 violent offenders imprisoned in Minnesota, 86 percent had had a pet sometime in their life that they considered special to them. Overall, 95 percent of the respondents valued pets for companionship, love, affection, protection and pleasure. Violent offenders were more likely to have a dog in their home while growing up. The control group had more animals as pets than dogs or cats, but the offender group had more "atypical" pets such as a baby tiger, cougar, and wolf pup. When asked what has happened to the special pet, over 60 percent of both groups lost their pets through death or theft; however, there were more pets that died of gunshots in the inmate group. In addition, the offender group tended to be more angry at the death of the pet. Strikingly, among the violent offenders, 80 percent wanted a dog or cat now as compared to 39 percent of the control group. This suggests something about the deprivation of the prison environment as well as the possibility of therapeutic intervention with pets among prison populations. Like the Kellert and Felthous study (1983), this study also found that most violent offenders had histories of extreme abuse as children (ten Bensel, Ward,

Kruttschnitt, Quigley and Anderson, 1984).

We also surveyed 206 teenagers between the ages of 13 and 18 living in two separate juvenile institutions and 32 youth living in an adolescent psychiatric ward in regard to their experiences with pets. We compared them to a control group of 269 youths from two urban public high schools. Of the 238 abused institutionalized youths we surveyed, 91 percent (N = 218) said that they had had a special pet and of these youths 99 percent said they either loved or liked their pets very much. Among our comparison group 90 percent (N = 242) had had a special pet and 97 percent said they either loved or liked their pet very much. This suggests that companion animals do indeed have a prominent place in the emotional lives of abused as well as non-abused children. It is also a corrective to those who suggest that pet ownership in itself will prevent emotional or behavioral disturbances in children. Merely having a special pet played no part in whether or not a child was eventually institutionalized (Robin, ten Bensel, Quigley and Anderson, 1983, 1984).

In considering the issue of abuse of animals, the authors found that the pets of the institutionalized group suffered more abuse; however, the abuser was usually someone other than the child. In a few instances, youths had to intervene against their parents to protect their pets. As one youth wrote, "He jumped on my bed and my mom beat him and I started yelling at her because she was hurting my dog." Another child wrote, "My dad and sister would hit and kick my cat sometimes because he would get mad when they teased him. I got mad and told them not to hurt him because he's helpless" (Robin, ten Bensel, Quigley and Anderson, 1983, 1984).

Of those youths who indicated that they mistreated their pets, sadness and remorse were the most common responses. For example, one child said, "I remember once I was punished for letting the dog out and so I hit him for that. I felt real bad after that and comforted it a lot. " All of those who mistreated their pets, except for one youth, indicated that they loved or liked their pets very much and felt bad about hurting their pets. Only one youth said he did not care that he hurt his pet. There was no self-reported evidence of sadism toward pets.

There were several instances of pets being harmed or killed as punishment to a child. According to Summit (1983), threatening to harm a child's pet is a common technique of child abusers to keep the child quiet about the abuse. In a recent child sexual abuse case discovered in a Los Angeles day care center, the adults involved allegedly silenced the children by butchering small animals in front of the children and threatening to do the same to their parents if they revealed the abuse. Mental health practitioners should routinely ask young people if anyone has ever hurt or threatened to hurt their animal.

Lenore Walker (1983) has suggested in her study on domestic violence that the best predictor of future violence was a history of past violent behavior. In her definition she included witnessing violent acts toward pets in the childhood home. At this point, without further studies, it is unclear what role, if any, violence toward pets plays in the emotional and behavioral disturbances of adolescents. Nonetheless, the abused institutionalized population experienced more violent pet loss than did the comparison group. They showed no evidence of callousness toward the sufferings of their pets and seemed to be troubled by the mistreatment of their pets.

CONCLUSION

Pets clearly play an important role in the lives of children. The relationship is characterized by deep feelings of love and care. It is enhanced by children's empathy toward the feeling of animals and their intuitive of having a common status with animals. As Freud (1853) wrote, "Children show no trace of arrogance which urges adult civilized men to draw a hard-and-fast line between their own nature and that of all other animals. Children have no scruples over allowing animals to rank as their full equals. Uninhibited as they are in the avowal of their bodily needs, they no doubt feel themselves more akin to animals than to their elders, who may well be a puzzle to them."

References

Anderson, R.K., Hart, B., and Hart, L. The pet connection: Its influence on our health and quality of life. Minneapolis: CENSHARE, 1984.

Beck, A., and Katcher, A.H. Between pets and people: The importance of animal companionship, New York: G.P. Putnam's Sons, 1983.

Bowen, M. Family psychotherapy with a schizophrenic in the hospital and in private practice. In I. Borzormenyi-Nagy and J.L. Framo (Eds.), Intensive family therapy. New York: Harper and Row, 1965.

Bowlby, J. Attachment and loss. In Attachment, Vol. I. London: Hogarth Press, 1969.

Brittain, R.P. The sadistic murderer. In Medicine, Science and the Law, 1970, 10:198-207.

Cain, A. A study of pets in the family system. In A. Katcher and A. Beck (Eds.) New perspectives on our lives with companion animals. Philadelphia: University of Pennsylvania Press, 1983.

Erickson, E. Identity and the life cycle. New York: W.W. Norton, 1980.

Feldmann, B.M. Why people own pets. In Animal Regulation Studies, 1977, 1: 87-94.

Felthous, A. Aggression against cats, dogs and people. In Child Psychiatry and Human Development, 1980, 10: 169-177.

Fogle, B. (ed.) Interrelations between people and pets Springfield, Illinois: Charles C. Thomas, 1981.

Fogle, B. Pets and their people. New York: The Viking Press, 1983.

Fox, M. Relationships between the human and non-human animals. In B. Fogle (Ed.) Interrelations between people and pets. Springfield, Illinois: Charles C. Thomas, 1981.

Freud, S. Letter to M. Bonaparte. In I. Simitis-Grubrich (Ed.), Sigmund Freud. New York: Harcourt Brace Jovanovich, 1976.

Freud, S. Totem and taboo. Standard edition. London: Hogarth Press and the Institute of Psychoanalysts, 1953, 1-161.

Fromm, E. The anatomy of human destructiveness. New York: Holt, Rinehart and Winston, 1973.

Hellman, D., and Blackman, N. Enuresis, firesetting and cruelty to animals: A triad predictive of adult crime. In American Journal of Psychiatry, 1966, 122: 1431-1435.

Katcher, A.H. Interactions between people and their pets: Form and function. In B. Fogle (Ed.), Interrelations between people and pets. Springfield, Illinois: Charles C. Thomas, 1981.

Katcher, A.H., and Beck, A. (Eds.) New perspectives on our lives with companion animals. Philadelphia: University of Pennsylvania Press, 1983.

Kellert, S., and Felthous, A. Childhood cruelty toward animals among criminals and non-criminals. Manuscript submitted for publication, 1983.

Kupferman, K. A latency boy's identity as a cat. In Psychoanalytic Study of the Child, 1977, 32: 192-215.

Levinson, B. Pet-oriented psychotherapy. Springfield, Illinois: Charles C. Thomas, 1969.

Levinson, B. The pet and child's bereavement. In Mental Hygiene, 1967, 51: 197-200.

Levinson, B. Pets, child development, and mental illness. In Journal of the American Veterinary Medical Association, 1970, 157: 1759-1766.

Levinson, B. Pets and human development. Springfield, Illinois: Charles C. Thomas, 1972.

Levinson, B. Pets and personality development. In Psychological Reports, 1978, 42: 1031-1038.

Lindsay, J. Hogarth: His art and his world. New York: Taplinger Publishing CO., 1979.

Mahler, M.S. On the first three subphases of the separation-individuation process. In International Journal of Psycho-Analysis, 1972, 53: 333-338.

May, R. Power and innocence, New York: W.W. Norton and Co., 1972.

Mead, M. Cultural factors in the cause of pathological homicide. In Bulletin of Menniger Clinic, 1964, 28: 11-22.

Midgley, M. Animals and why they matter. Athens: University of Georgia Press, 1984.

Montaigne, M. de. The essays of Montaigne. New York: Oxford University Press, 1953.

Morris, D. The naked ape. New York: McGraw-Hill, 1967.

Nieberg, H.A., and Fischer, A. Pet loss: A thoughtful guide for adults and children. New York: Harper and Row, 1982.

Perin, C. Dogs as symbols in human development. In B. Fogle (Ed.), Interrelations between people and pets. Springfield, Illinois: Charles C. Thomas, 1981.

Robin, M., ten Bensel, R.W., Quigley, J., and Anderson, R.K. Childhood pets and the psychosocial development of adolescents. In A. Katcher and A. Beck (Eds.), New perspectives on our lives with companion animals. Philadelphia: University of Pennsylvania Press, 1983.

Robin, M., ten Bensel, R.W., Quigley, J., and Anderson, R.K. Abused children and their pet animals. In R.K. Anderson, B. Hart, and L. Hart (Eds.) The pet connection: Its influence on our health and the quality of life. Minneapolis: CENSHARE, 1984.

Ruby, J. Images of the family: The symbolic implications of animal photography. In A. Katcher and A. Beck (Eds.), New perspectives on our lives with companion animals. Philadelphia: University of Pennsylvania Press, 1983.

Rynearson, E.K. Humans and pets and attachment. In British Journal of Psychiatry, 1978, 133: 550-555.

Schowalter, J.E. The use and abuse of pets. In Journal of the American Academy Child Psychiatry, 1983, 22: 68-72.

Searles, H.F. The non-human environment. New York: International University Press, 1960.

Sherick, I. The significance of pets for children. In Psychoanalytic Study of the Child, 1981, 36: 193-215.

Siegel, A. Reaching severely withdrawn through pet therapy. In American Journal of Psychiatry, 1962, 118: 1045-1046.

Stewart, M. Loss of a pet - loss of a person: A comparative study of bereavement. In A. Katcher and A. Beck (Eds.) New perspectives on our lives with companion animals. Philadelphia: University of Pennsylvania Press, 1093.

Summit, R. The child sexual abuse accommodation syndrome. In Child Abuse and Neglect, 1983, 7: 181.

Tapia, F. Children who are cruel to animals. In Child Psychiatry and Human Development, 1971, 2: 70-77.

ten Bensel, R.W. Historical perspectives on human values for animals and vulnerable people. In R.K. Anderson, B. Hart, and L. Hart (Eds.), The pet connection: Its influence on our health and quality of life. Minneapolis: CENSHARE, 1984.

ten Bensel, R.W., Ward, D.A., Kruttschnitt, C., Quigley, J., and Anderson, R.K. Attitudes of Violent criminals towards animals. In R.K. Anderson, B. Hart, and L. Hart (Eds.), The pet connection: Its influence on our health and quality of life. Minneapolis: CENSHARE, 1984.

Thomas, K. Man and the natural world. New York: Pantheon Books, 1983.

Walker, K. The battered women syndrome story. In D. Finkelhor (Ed.), The dark side of families. Beverly Hills: Sage Productions, 1983.

Michael Robin and Robert ten Bensel are at the University of Minnesota, School of Public Health, Program in Maternal and Child Health, Box 197, Mayo Memorial Building, 420 Delaware Street, S.E., Minneapolis, MN 55455.

The above article appeared in the journal Marriage and Family Review, pages 63-78. 1985 by The Hayworth Press, Inc., 10 Alice Street, Binghamton, NY 13904, and is reprinted here with permission.

The Latham Letter, Vol. XI, No. 2, Spring, 1990, pp. 1, 17-23.

THE ENVIRONMENT

Population Growth and the Environment

Population-Environment Balance's position is based on the realization that a stable U.S. population size is essential if we are to prevent further deterioration of the very system that supports us -- our environment and natural resource base. Regardless of how conservatively we use resources, the fundamental fact is that growing numbers of people unavoidably place increasing demands on our natural and social environment. More people mean more energy use, more traffic jams, more production of toxic wastes and increased tensions which result from living in crowded urban environments. However efficient we may be in use of resources and however much we conserve in our attempt to preserve our environment, more people simply mean more stress on the environment as the phenomena of crowding, deforestation, acid rain, global warming and the whole litany of environmental ills amply demonstrates.

WATER AVAILABILITY AND POPULATION GROWTH

The United States is facing grave water shortages, particularly in those areas of the country with the highest population growth. Groundwater is being pumped out of the ground faster than it is being replaced in 35 of the 48 contiguous states. This depletion, often called "water mining," involves both localized groundwater declines at particular sites scattered across the country, as well as regionally significant groundwater declines in three major areas: southern Arizona, the High Plains (a wide swath extending across several states and running from Montana to Texas, and California. The groundwater overdraft is occurring because agricultural, industrial, and residential demand is outstripping the renewable surface and groundwater supply.

Our national supply of renewable freshwater is approximately 1,400 billion gallons per day, about three times our daily national withdrawal rate. In view of this overall abundance, it may seem strange that parts of the country are experiencing water shortages. But many areas are experiencing such shortages because of the uneven distribution of water supply and water demand. The eastern part of the country (roughly, the area to the east of the Mississippi River) has 37 percent of the nation's renewable water supply, but only accounts for 8 percent of the nation's demand. In contrast, the southwestern area of the country has only 6 percent of the nation's renewable water supply, but its large farms and sprawling urban areas account for fully 31 percent of the nation's demands.

As human population continues to grow, particularly in the dry southwest, state and federal governments are being pressured to implement large-scale water projects to import water needed to meet the increasing demand. The cost of imported water is usually much greater than the price that people who use the water actually pay for it. Not only do artificially low water prices encourage wasteful use by current users, but low water prices can encourage new users to move into the area. Once more people have migrated to arid regions than local water resources can support, governments come under political pressure to provide adequate supplies of cheap water, even if it means taking water away from other regions and charging people who do not use the water for the costs of importing it.

POPULATION GROWTH AND FARM LOSS

A recent study of the top 20 percent of agricultural counties in each state -- the High Market Value Farming Counties -- found that over half of the (58 percent) were either inside or adjacent to metropolitan areas. Thirty percent of the value of America's total agricultural production comes from the High Market Value Farming Counties located on the rapidly urbanizing fringe of America's metropolitan areas. The population of these counties grew by 20 percent during the 1970s, and metropolitan growth is accelerating. Unless measures are taken to soon slow population growth and control urban expansion through public direction of land use, much of this irreplaceable farmland will be gone forever.

As urban areas expand, farmland is consumed by houses, office buildings, shopping malls, and parking lots. Prime farmland is usually level, stable, and well-drained -- characteristics that make the land well-suited for development.

Internal migration and immigration from other countries are sources of the rapid growth in America's South and West. California, Texas and Florida -- three extremely important agricultural states -- each grew by over a million people between 1980 and 1984. This population growth is threatening the nation's most valuable farmland. California is losing 44,000 acres of cropland a year to urban uses. Texas lost over 600,000 acres of prime farmland to permanent non-agricultural uses between 1977 and 1982. The National Agricultural Lands Study projects that at the current rate of conversion, Florida will lose all its prime agricultural land by the year 2000.

EXCESS IMMIGRATION AND THE ENVIRONMENT

The United States population is increasing by nearly 3 million per year. Since immigration from foreign countries causes over 40% of U.S. population growth (and nearly 60% of the population growth of some states such as Florida and California), and since the U.S., too, has a limit on its carrying capacity (the number of people who can be sustainably supported in a given area without degrading the natural, social, cultural, and economic environment for future generations), excess immigration creates a significant environmental threat.

Worldwide, a common response to carrying capacity problems is to migrate to areas where the carrying capacity has not yet been pushed beyond the limit or is perceived to still provide opportunities. But the problem is that such migration not only threatens the carrying capacity of the destination countries, but also creates the harmful illusion that continued population growth is an acceptable option.

Populations try to move out of countries where they have overwhelmed the carrying capacity. Today, the pressures from every continent continue to increase -- world population is growing by 97 million per year! Many have already come to the United States, but no region, including the United States, has the carrying capacity to absorb all those desiring to immigrate. It is doubly unfortunate, therefore, that the perception of opportunity in the U.S. acts as a disincentive for over-crowded countries to face and begin to correct overpopulation problems at home.

Allowing too much immigration both creates an environmental threat and sends a misleading signal. Perhaps all countries should consider limiting immigration to levels within their carrying capacities in order to more effectively protect the environment. Allowing immigration in excess of carrying capacity ignores the limits in both the sending and receiving countries. Such a disregard represents a serious threat to the environments of all countries involved.

THE ULTIMATE ENVIRONMENTAL THREAT: OVERPOPULATION

One result of overpopulation is that resources are depleted and the environment of degraded to the point that an area loses part of its capacity to support population in the future. When the carrying capacity is exceeded, the environmental damage is usually so severe that the population carrying capacity for future generations is greatly reduced.

The point is simple enough: more people demand more of the shrinking resources, and, using them, create more pollution. Global warming, species extinction, acid rain, deforestation of the Tsongas and other national forests are among the signals that the United States' and the world's population increase is pushing the environment beyond its ability to sustain a desirable quality of life.

The above information was provided by Population-Environment Balance. For further information, write: Population-Environment Balance, 1325 G Street, N.W.m, Suite 1003, Washington, DC 20005.

The Latham Letter, Vol XII, No. 2, Spring 1991, pp. 1, 22.

Environmental Protection for the 1990s and Beyond

Milton Russell

Bob Dylan's "The Times They Are A-Changing" was a theme song for a generation that heralded the environmental awakening in the United States of two decades ago. Given the point at which we find ourselves with many environmental questions, it might well be an appropriate anthem again. Four basic changes are taking place now, but will be even more important in the future. The first change is in types of environmental problems that will absorb most of our talents and efforts. In brief, the environmental enterprise will become more concerned about toxic substances and more conscious of ecological impacts that often accompany and sometimes outweigh public health problems.

The second change is in the targets of efforts to improve the environment. The importance of changing the behavior of a few industrial polluters will decline, to be replaced by the need to alter the ways in which literally millions of private citizens go about their daily lives. Third, the role of environmental professionals will shift, and the tools with which they traditionally sought beneficial environmental results will have to be augmented to reach those new targets. Finally, there will be a profound change in the roles of the federal government on the one hand, and of states, communities, and private citizens on the other.

To understand these changes, it is useful to look back and see where the country has been with regard to the environment, and to assess where it is now. There are features in this experience much different from those that molded the European reaction to environmental concerns. Better understanding of these differences may improve the ability of countries to borrow wisely from each other in meeting the environmental challenges we face.

HOW WE GOT HERE

As Calvin Coolidge put it, "The business of America is business,"[1] and for most of our history that has been true. That business meant converting a virgin land to the most productive economy ever known. The national treasure awaiting exploitation included not only the raw materials needed for industrial supremacy, but also the natural sinks of great rivers and lakes, tidal bays, seas, vast skies, and, most of all, broad sketches of land in which to dispose cheaply of the waste products of industry and of life itself. The nest we fouled was indeed our own. But it was enormous, and for those who did not choose to push on to fresh frontiers, the stench was overpowered by the stronger perfume of money.

Pockets of despoliation were always present. They showed up first in the industrial East, where, for example, salmon disappeared from some streams early in the 19th century.[2] They existed in the coal regions where a combination of strip mining and acid drainage turned productive streams into silt laden, lifeless sewers sacrificed to America's industrial juggernaut. The pockets showed up as choking, lung-searing air in our richest cities. But there were still many other streams and lakes, and at least those better off could escape the foul air by repairing to the surrounding suburbs. It is sobering to reflect on a near ubiquitous institution that began a couple of generations ago -- the "fresh air funds" that took poor children from the city. "Fresh air" was not a cure name for a life enriching experience; fresh air was the experience, and the goal.

In short, the vastness of our country sustained the reality that there was an "away" to which pollution could be shunted, that ecological insults did not threaten the fabric of natural life on which we all depended and which we valued; and that a nest too fouled could be abandoned, to start afresh.

The historian Frederick Jackson Turner marked the end of the first phase of American life with the closing of the geographic frontier around 1890. His thesis was that the closing of that safety valve brought a profound change in the way Americans thought of themselves, their country, and the rest of the world.[3] It took decades for the full implications of a closed geographic frontier to sink in and to change the American outlook. We have only recently come to realize that our environmental frontier has likewise closed, and we are still sorting out the meaning for the way we behave and relate to each other and natural systems.

The environmental frontier -- the sense that there is an "away" -- sealed up when Americans began to see that the pustules of environmental degradation were no longer isolated, but instead threatened the land with a consuming rash. That recognition was a long time coming. It was presaged by the recurrent motif of environmental disaster found in post-World War II science fiction. The popularity of John Kenneth Galbraith's The Affluent Society in the 1950s suggests that people identified with his outrage at expansive private consumption in the presence of public squalor.[4] Silent Spring caused many to question whether progress was a bargain when bought at the cost of birthright pleasures such as the song of a bird in the wild.[5] The Donora disaster of 1948, where 20 people died and almost half a town was stricken with illness from polluted air, altered forever the vision of belching smokestacks as an acceptable nuisance.[6] Buckminster Fuller's compelling image of Spaceship Earth dramatized that there really was no "away" -- that "here" was all there was. The flaming

Cuyahoga River emptying into the dying Lake Erie was a graphic reminder that the waste sink of the industrial Midwest was finally spilling over.

The realization dawned that Mother Earth was being abused by her children. She was reaching the point where she could no longer absorb the punishment and benignly forgive. That realization was given form and substance by Earth Day in 1970. Millions of Americans found that their concern and outrage were shared by others. No longer was environmental degradation a local issue or the province of an alienated few, but a national issue that engaged the energy and dedication of Main Street America.[7]

With congressional elections impending and a presidential election on the horizon in 1972, suddenly the environment was a hot political issue. Scarcely two months after Earth Day, President Richard Nixon proposed to Congress the establishment of the Environmental Protection Agency (EPA) as an umbrella organization to manage the national effort to heal the environment.[8]

Although industry was largely skeptical of this seeming fad, political and environmental leaders were aware of deep grassroots support and used that support to launch and maintain a decade-long surge of environmental activism. That support has been maintained and even grown.[9] Congress passed several sweeping new statutes in quick succession, giving EPA the power and responsibility to reverse the worst excesses of pollution.

The problems were gross and the solutions appropriately crude. There was little need and no patience for careful science, discriminating regulations, or a weighing of the costs. The mission was to blanket the nation with a consistent set of requirements that would keep major pollutant streams out of the air and water. Industrial polluters were the obvious target. Theirs were the largest and most visible waste streams and the most susceptible to tough-and-ready control.

Environmental Protection for the 1990's and Beyond 209

Despite the enormous odds against it the hell-bent-for-leather strategy worked quite well on the whole. The environment is not only cleaner now than it was then; it is much cleaner than it would have been had the nation not responded. For example, air quality is generally better now than in 1970 despite a growth in gross national product of 50 percent and an increase of 17 percent in population.[10] As a dramatic example, ambient lead concentrations in urban areas dropped by 79 percent from 1976 to 1985.[11] Most of our rivers are cleaner now, or are at least holding their own.[12] Private activism and government regulation unquestionably have turned the advancing tide of the most visible and repulsive pollution.

So much is history. Through those efforts major environmental risks to health have been substantially reduced. People live longer; they have safer air, water and food; and they enjoy more pleasing surroundings than before.

But the focus here is on the future, and on the job yet undone. The sobering evidence is that this country will not be able to continue environmental progress and achieve its goals by simply maintaining or even intensifying current efforts. The challenges for the 1990s are fundamentally different from those of 1970, and new approaches will be necessary to meet them and to maintain the progress already achieved. At the same time, no backsliding can be permitted on past progress. The premise of what follows is that the baseline of success from existing efforts will be maintained.

CHANGE IN ENVIRONMENTAL THREATS

The perceived threats to environmental quality have now changed as compared with those of a generation ago. Then one could see, smell, and taste the problem to be attacked. When the battle was going well, human senses knew it. Now attention has shifted to toxic chemicals, some of which can pose subtle threats at concentrations almost mystically small, discernable only through advanced technology. And toxic chemicals are everywhere -- in the ambient air, in the home, in the food chain, and in the water. Unlike the case with smoke or sewage, however, to find a toxic chemical is not to define a course of action. There may be a hazard -- that is, a chemical may be intrinsically harmful -- but unless it is harmful at relevant doses and unless people or ecological systems are exposed at those levels, there is no risk. It is the amount of risk and what can be done about it that are important aspects in guiding action.

The problem of toxic chemicals is exacerbated when suspected carcinogens are involved. For the pollutants of major consequence in the 1970s, there were natural stopping places for control because, below some concentrations, natural systems were capable of absorbing the insult by rendering the pollutant harmless. Similarly, for many of these pollutants, human health thresholds exist so that few if any harmful effects are observed below a certain level. For most carcinogens, however, current scientific theory holds that any dose can be assumed to have some probability of harm, although perhaps very small. Zero risk does not imply zero smoke in the air or zero sewage in the water, but it does imply zero exposure to most carcinogens.[13]

Yet, it is a delusion to set a goal of eliminating toxic chemicals in the same way that the goal of "swimmable, fishable" water was set. That is a prescription of frustration. There are not the scientists, engineers, and technicians to study every chemical and devise controls to eliminate all risks. And if there were, and even if it were physically possible, there is not the social and political will to pay the price in reduced consumption of goods and services that zero risk would entail. Therefore, added to the technical enterprise of reducing toxic pollution is a demanding social and political task. That task is to select from a near infinity of potential targets which toxic chemicals should get priority attention and to determine how far to go in controlling those selected.[14]

Making these choices is information-intensive. In addition, it demands the highest level of leadership and public involvement. This is because the public will be faced with explicit tradeoffs between reducing toxic risks still further and reducing yet more personal freedom and the consumption of other goods and services -- including risk reductions elsewhere. Such choices are inherently cruel, and the temptation to simply turn them aside is great. Indeed, the political system has mostly succeeded until now in casting a rhetorical veil between such choices and the public when it comes to the environment. It has done so by promising progress toward perfection, by asserting that the cost would be paid by others, or disingenuously, by simply fuzzing or denying the tradeoffs presented. But that time is passing.

Another change in focus is toward greater concerns with protection of ecological systems. Protection of health was understandably the early priority. Fortunately, many of the actions of the first generation of pollution controls protected natural systems as well. It has become clear, however, that this is not enough to assure that future citizens will enjoy the birds, the fish, the diversity of plants, and the vistas this generation enjoys. Indeed, even within this generation subtle but significant losses have occurred.

Public attention has been drawn to the plight of endangered species such as the California condor, the blackfooted ferret, and the dusky seaside sparrow, the latter of which apparently became extinct in June 1987. But more telling, perhaps, has been the almost imperceptible draining away of habitat and the subsequent diminution in size and diversity of animal and plant populations. It is striking to read accounts such as Edwin Way Teale's classic Autumn Across America, based on observations in the late 1940s.[15] He depicts as commonplace flights of birds in places where they are now nowhere to be found, at least in the numbers he described. Not that success stories are lacking, as the expansion and extension of the wild turkey population demonstrated strikingly. The growth of white-tailed deer populations and the expansion of the coyote range are also notable.

Captive breeding programs and intensive management have brought back a sustainable population of whooping cranes. The bald eagle population is growing and extending its range as DDT is purged from the system, and through active intervention to protect nesting sites and to transplant young birds to new areas. Protection of special habitats has preserved, at least for now, the grizzly and the wolf. But the common vision of the ecological legacy we want to leave is not one reduced to open zoos and wildlife preserves, as important as they are, but of healthy productive ecological communities, thriving along with a human population in a common setting. To achieve that legacy will take action -- to protect the wetlands, to preserve greenbelts, to improve water quality, and to avoid ecological insults even beyond those injurious to human health.

CHANGE IN SUBJECTS TO CONTROL

When it comes to formulating the next generation of environmental controls, environmental professionals will soon learn sympathy for General Charles Cornwallis. He was trained to fight set battles against neat columns of brightly dressed soldiers, only to find himself peppered with fire from colonial rebels behind every tree in the woods.

In the old days, the polluter was big industry, easy to identify, easy to paint as the culprit, and relatively easy to control. Perhaps as important, it was easy to pass off the cost of cleaning up as someelse's problem, although in truth it was borne by all of us as workers, investors, consumers, and taxpayers. For the future, however, much of the remaining pollution that will cause the most risk is from widely dispersed sources whose control will depend on changing the behavior of individual citizens.[16] Here the cost in dollars, inconvenience, and lost amenities cannot be passed off or hidden.

People want clean air, in general, but few want the nuisance or expense of an annual automobile inspection. Still less do they want to reduce driving or give up favored consumer goods. Toxic air emissions from industry are pretty well controlled; the evidence is that the most risk now comes from individually small but cumulatively large sources, such as dry cleaners and woodstoves.[17] The air indoors is more hazardous than that outside, but homeowners want clean ovens, painted walls, and inexpensive warmth in the winter. Today's tight home construction that brings down fuel bills ironically exacerbates risks from everything from smoking to naturally occurring radon.

Modern sewage treatment plants mean sharply higher, and very unpopular, monthly household bills, especially now that the federal government will be paying less of the cost under the Water Quality Act of 1987.[18] The major threat to wetlands is not from large conversions sponsored by big corporations but from small changes imposed by householders who want to bulkhead their shorelines or farmers who want to drain and plow a prairie pothole or convert hardwood bottomlands to soybean production.[19] Rivers and lakes are threatened by run-off from streets, farms, and forests and by the do-it-yourself motorist who dumps used oil down the storm sewer.[20] People want streams and estuaries to support fish, game, and recreation, but they also want shirts clean, and so the phosphorous continues to flow into the water. They want industrial hazardous waste properly managed, but few support a facility to do just that if there is a chance that it might be located in their neighborhood.

The litany goes on. The problems are obvious; the solutions are commensurately murky. Certainly they are not the traditional ones. The tool kit of environmental protection that got us this far does not have the wrenches and levers to fix the remaining problems. New tools are needed, and environmental professionals will have to learn to use them.

CHANGE IN ROLES

Devising and applying those tools represents the third change needed for effective environmental protection for the 1990s and beyond. Here the waters are scarcely charted, but the general course appears clear.

First of all, the people who are to provide the solution must understand and accept the problem as theirs. Unless the health risks are understood and judged unacceptable, it is fruitless to expect behavior to change. In this new phase of environmental progress, action comes only when the polluters choose to impose change on themselves and their fellows. Those who are part of the problem are also those who must agree upon and carry out the solution. This is not a situation amenable to command and control; it is one that demands coalition and consensus.

In this new era, the role of the environmental professional is one of assessing risk -- defining the nature, scope, and magnitude of the problem. It is also one of communicating that risk assessment to the affected public and of laying out possible fixes to the problems discovered. Then an informed community can make its choice and find a way to bear fairly the burdens any actions impose. The risks identified can be weighed against the cost of reducing them, and a balance struck based on prevailing values.

Individuals will be making the risk management decisions for themselves and with their fellows, for each other. Questions such as these will be posed: Is avoiding that herbicide worth tolerating dandelions in the neighborhood lawns? Do I want to test my home for radon? If I find it, will I pay to cut my risk of lung cancer? Should I put a catalytic converter on my woodstove to avoid harmful emissions in my house -- and how much pressure will I withstand from my neighbors who don't want those emissions in their back yard? Will I support land use control to protect wetlands if that means that a marina is further away or that the longed-for beach house cannot be built on the already-purchased lot? Or that jobs will

not be created here because lack of wharfage means that a plant will be located elsewhere? Will I vote in or out a mayor who supports an incinerator to transform my garbage safely, knowing that the facility may be sited in my neighborhood?

Environmental professionals who have joyfully worn the cloak of philosopher kings, forcing some people to clean up their act for the good of others, will have to change their clothes. The new message is: We are all in this together. Here are the facts. What should we do about it?

In brief, this means opening up the process of environmental decision to those who must bear the consequences and pay the costs. And then accepting their judgment. This is a new way of doing things, and for those steeped in the past and concerned about the future, it may seem dangerous. But Thomas Jefferson did not think so. He said:

I know of no safe depository of the ultimate powers of the society but the people themselves; and if we think them not enlightened enough to exercise their control with a wholesome discretion, the remedy is not to take it from them, but to inform their discretion.21

Besides, when it comes to most of the environmental problems left on our plate, there is no realistic option except to share control.

CHANGE IN INSTITUTIONS

Three types of change have been addressed so far: change in types of environmental problems, change in who causes them, and change in how they must be fixed.

The final change must come in the institutions by which our society achieves and assures environmental quality. The roles and relationships among the parties that have been traditionally entrusted with environmental protection -- the federal government, the states, the communities, the private sector -- have changed and will change still more.

In many respects states are better equipped than EPA to carry out the everyday work of environmental protection. Twenty years of building have brought this country to the point where most states have competent, highly professional institutions to ensure consistent and responsible environmental protection within their own borders. A few years ago, EPA acknowledged that fact and issued a new policy on state and federal roles.22

That policy, negotiated between EPA and state participants, defined a new partnership for environmental protection, one that assigned to each party the functions it is best fit to perform. Briefly, states take on the lion's share of responsibility for field-level program operations, such as permitting, inspecting, and enforcing. EPA, on the other hand, acts as a sort of national franchiser -- an environmental McDonalds. EPA has principal responsibility to provide national leadership, evaluation, and support to state environmental programs; to issue national standards and regulations; to undertake research and information collection; to back up states on the odd occasion when they cannot perform; and to represent the needs of both partners before Congress and within the administration. This system was well conceived and it is essential that it continue to provide the national framework and baseline for environmental protection.

This partnership, however, has far to go to reach success in treating even traditional environmental concerns. For EPA and Congress, which prescribes its behavior through statutes, the realization that the states and localities must be trusted, or built to a position of trust, requires a leap of faith that many have been unable to make. Just as telling, the states and localities are sometimes reluctant to accept responsibility for tough actions when they can turn to their less locally constrained Uncle Sam to play the gorilla in the closet. It is a lot easier for a local or state official to hold voters to a federally mandated standard than to achieve the same goals by getting local legislation passed and then by enforcing this close-to-home-grown decision.

The points outlined above suggest, however, that these ongoing institutions must also be enriched to meet the challenges of the future. The direction of the needed change seems clear: state, communities, and citizens will require organizations to diagnose local problems, to find acceptable solutions, and to induce behavioral change. And on the other side, EPA will need to back off on the micro-management that places a straitjacket on those who need to deal with their special problems in ways they find acceptable. Finally Congress will need to provide flexibility in the statutes to allow states and communities more freedom to determine for themselves at least the "how" and in some cases the "what" and the "when." Again, none of these changes will be easy. But some are already underway.

States are adjusting to the new realities by rethinking their approach to environmental planning. Several states have joined with EPA in developing new ways of diagnosing their specific multi-media environmental problems and in planning for their correction. In Maryland, for example, a state task force has identified 20 of the most troublesome environmental issues, such as indoor air pollution and groundwater contamination, and held workshops on each to define appropriate goals and recommendations for action. While the process is being carried out by state agencies, there has been extensive involvement by citizen experts and other nongovernment participants. In Delaware the governor has organized a broadly representative Environmental Legacy Program to define a desired environmental future and design an action program to ensure it. In Pennsylvania state government and private citizens have joined the federal EPA in an effort to determine where the most risks are and to establish priorities in addressing them.

The role of communities is expanding as well. EPA has been working with several localities in providing technical support for their evaluation of complex environmental conditions at the local level. These "Integrated Environmental Management Projects" have been set up in places as diverse as Philadelphia; Baltimore; Santa Clara, California; West

Virginia's Kanawha Valley; and most recently, in Denver. Although each project has been different, all of them have organized local interests in the collection of empirical information on pollution, its location, its attendant risks, and the options available for reducing those risks. The focus is on what risks are left after the safety net of national controls is imposed.23

Based on quantitative risk assessments when possible and qualitative ones otherwise, these local communities are coming up with priorities and plans for action. The data and technical assistance provided to the community by EPA empowers local residents to decide which risks are acceptable and which are not.

In other places, grassroots movements have grown up to meet environmental needs. One instance involves Oregon's Tillamook Bay, an area famous for both oysters and cheese. Coastal Oregon gets about 100 inches of rain a year. The attendant run-off washes dairy cattle wastes into the bay in such volumes that in 1977 the Food and Drug Administration threatened to close the bay to oyster harvesting under a little-used federal regulation that would have replaced state authority. To preserve their local environment and economy, local oystermen and dairy operators have voluntarily joined to institute best available management practices at over half the area's dairies. They sought and received help from the state, the Soil Conservation District, the Department of Agriculture, and EPA. But the solution was initialed and implemented by local citizens and industry. Oysters and cheese production now happily coexist, and local institutions see to it that neighbors do not impose on one another.24

In short, there is a world of creativity in our communities that our previous over-reliance on federal regulation has left untapped. Federal regulations are powerful but limited tools -- like the driver in a golf bag that has its place at the tee but not on the putting green. States, localities, and private citizens have at their command many more subtle devices that are properly denied at the federal level. Prudent zoning and other land-use controls come to mind as particularly incisive tools for handling specific problems. In an increasing number of environmental situations, EPA would do best to stay at the tee while others more effectively play on the green.

WHERE ARE WE GOING

Daniel Boone wanted to move west when he could smell another's woodsmoke. He could. We cannot. The environmental frontier is closed. With no "away," the output and consumption that meets people's wants, and the dark underside of attendant environmental risks, must be managed in a closed system.

For example, pesticides and herbicides will continue to be needed but more careful attention to the way they are used and to their true benefits and environmental dangers can lead to the use of safer, more selective, and less persistent ones and to less of them being released. Hazardous waste will continue to be produced, but its volume can be reduced, more of it can be destroyed or recycled. and the risk from the remainder can be lessened by selecting wisely where it finally comes to rest. Or again, pressing social wants will lead to destruction of habitat, but there will be less destruction when ecological damages are put into the equation, and much of the loss can be offset by proper attention to mitigation and enhancement of habitat elsewhere. The issue is balance. The questions to be posed are how our constrained space and limited resources are to be used, for what ends, and who is to decide.

Great progress has been made under the statutes, and institutions that grew up over the past generation, and that hard-won progress must be protected. Yet the seemingly well-marked though arduous road to environmental quality that has been followed must be rerouted to deal with new realities:

The illusion that safe can mean risk-free has been dispelled with better understanding of cancer and by the ubiquity of toxic chemicals that are essential to daily life.

The recognition that protecting people does not automatically protect ecological systems came as we discovered that power plant emissions that meet health standards are implicated in reducing the biological diversity of lakes.

The belief that it was others who had to clean up has been replaced by the reality that "the enemy is us," even as we sort our garbage for the incinerator and take our cars to have their emissions checked.

The view that solutions could be devised, enunciated, and policed from on high has evaporated as communities wrestle with siting hazardous waste facilities and choosing how to cut down on the hydrocarbon emissions that create local smog.

The reflective posture of "leave it to the feds" has come up against the diversity of problems, the complexity of local solutions, and the need for outcomes that communities can accept as fair.

The easy part of meeting the environmental challenge of the 1990s will be scientific and technical. This society knows how to mobilize and use its talent to gain the knowledge it needs. All it takes is the will to devote the needed resources to the task, and time. The hard part lies in accomplishing the shifts in attitude, behavior, and institutions needed to comport with this stage of the American quest for a safe, healthy, ecologically secure environment.

Bob Dylan was the troubadour of a great period of spontaneous change in America. But as he noted, the times are changing indeed. Today's environmental professionals will not be true to their mission unless they take a leadership role in fostering the daunting but essential changes that will mold an environmental legacy they will be proud to leave to the 21st century.

Notes

1. Calvin Coolidge (Speech to the American Society of Newspaper Editors, 17 January 1925).

2. A. Netboy, The Salmon: Their Fight for Survival (Boston: Houghton Mifflin Co., 1974). 169-82.

3. F.J. Turner, Frontier in American History (New York: Holt, Rinehart, & Winston, Inc, 1920).

4. J.K. Galbraith, The Affluent Society (Boston: Houghton Mifflin Co., 1958).

5. R.L. Carson, Silent Spring (Boston: Houghton Mifflin Co., 1962).

6 "Delayed Effects of Smog Studied," New York Times, 30 November 1958.

7. "Nation Set to Observe Earth Day," New York Times, 21 April 1970.

8. R. Nixon, "Message to Congress Transmitting Reorganization Plan 3 of 1970: Environmental Protection Agency," Public Papers of the President (Washington DC: U.S. Government Printing Office [GPO], 1970).

9. See, for example, Riley E. Dunlap, "Public Opinion in the Environment in the Reagan Era," Environment, July/August 1987.

10. U.S. Environmental Protection Agency, "National Air Quality and Emissions Trends Report, 1985" (Washington DC, 1985), 1-5 to 1-19: Council of Economic Advisers. Economic Report of the President, 1987 (Washington DC: GPO, 1987), 246, 279.

11. EPA, note 10 above, 3-38.

12. EPA, National Water Quality Inventory: 1984 Report to Congress. EPA 440/4-85-029 (Washington DC, EPA 1985).

13. 51 Federal Register, 33992-34003 (1986).

14. EPA, "Risk Assessment and Management: Framework for Decision Making" (Washington DC: December 1984).

15. E.W. Teale, Autumn Across America (New York: Dodd Mead, 1956).

16. EPA, "Unfinished Business: A Comparative Assessment of Environmental Problems, Overview Report" (Washington DC, 1987).

17. EPA Office of Air and Radiation, Office of Policy, Planning and Evaluation. "The Air Toxics Problem in the United States: An Analysis of Cancer Risks for Selected Pollutants" EPA 450/1-85-001 (Washington DC. 1985).

18. Pub. L. No 100-4 (1987)

19. William E. Odum, "Environmental Degradation and the Tyranny of Small Decisions" BioScience 32 (1982): 728-29.

20. EPA Office of Water Program Operations, Report to Congress: Nonpoint Source Pollution in the U.S. (Washington DC: EPA, 1984).

21. T. Jefferson, Letter to William Charles Jarvis, 28 September 1820.

22. "EPA Policy Concerning Delegation to State and Local Governments" internal policy document, 4 April 1984.

23. EPA, "Baltimore Integrated Environmental Management Project: Phase 1 Report" (Washington DC May 1987): EPA "Final Report of the Philadelphia Integrated Environmental Management Project" (Washington DC December 1986); EPA, "Draft Kanawha Valley, West Virginia Toxic

Screening Study Report" (Washington, DC February 1987); E. Haemiseggar, A. Jones, and F. Reinhardt, "EPA's Experiences with Assessment of Site-Specific Environmental Problems: A Review of IEMP's Geographic Study of Philadelphia," Journal of the Air Pollution Control Association 35 (1985): 809-15.

24. John E. Jackson, "Shellfish Sanitation in Oregon: Can it Be Achieved Through Pollution Source Management?" in EPA Office of Regulations and Standards, "Perspectives on Nonpoint Source Pollution," EPA 440/5-85-001 (Washington DC, 1985).

Milton Russell was assistant administrator for Policy, Planning and Evaluation at the U.S. Environmental Protection Agency from 1984 until last spring [spring, 1987]. He now has a joint appointment at the University of Tennessee in Knoxville, where he is a professor of economics and Senior Fellow in both Energy, Environment and Resource Center and at the Institute for Waste Management and Education, and at Oak Ridge National Laboratory, where he is a senior economist.

The author acknowledges the assistance of Thomas Kelly in the preparation of a previous version of this paper.

Republished from Environment, Vol 29, No. 7, pages 12-15 & 34-38, September 1987. Reprinted with permission of the Helen Dwight Reid Educational Foundation. Published by Heldref Publications 4000 Albemarle St., N.W., Washington D.C. 20016. Copyright © 1987.

The Latham Letter, Vol. IX, No. 1, Winter 1987/88, p. 1-7.

PHILOSOPHY

The Evolution of Animals in Moral Philosophy

Steve F. Sapontzis, Ph.D.

In mainstream Western moral philosophy, animals have passed through one stage, are currently in a second, and, if the animal liberation movement is successful, will be entering a third. In the first of these stages, animals were excluded from being direct objects of moral concern at all. In the second stage, animals have become objects of compassion and of the moral concerns that cluster about compassion. However, they remain resources to be exploited for human benefit. In the third stage, animals will become objects of our concern with fairness and the moral concepts which cluster about the idea of justice. I want to go over these stages with you, indicating how they originated and developed and detailing the differences among them.

STAGE I: NATURAL RESOURCES

The classical concepts of animals which have contributed to Western moral tradition can be divided into those emanating from the Biblical book of Genesis and those emanating from the Greek philosopher Aristotle. While Greek philosophy and Biblical teaching differ in many and significant ways, they share two ideas which have been crucial in shaping our traditional attitudes toward animals. The first of these is the belief that nature is ordered hierarchically, with human beings, or at least some human beings, at the apex of creation. The second of these shared ideas is the belief that purpose is a fundamental category in nature, with the lower orders of nature having been created for the purpose of fulfilling the needs of the higher orders.

In the story of creation contained in the first chapter of Genesis, it is said that people were created in the "image" of God and that we were given "dominion" over the rest of creation. Being the sole image of God in creation provides us a unique status in the universe, and being entrusted with dominion over God's creation is recognition of this special position. One might expect that the animals would have benefitted handsomely from those metaphors. After all, God is supposed to be a loving parent, who is solicitous of those He has created. If humans are the image of God, then they, too, should be concerned to love and cherish, aid and protect what God has created. Furthermore, God has, according to Genesis, made us His vice-regent on earth; He has put us here to rule and administer what He has created and called "good." You'd think that a subordinate given this awesome responsibility by an all-powerful Creator of the Universe would want to be very careful that he/she did nothing to harm or detract from God's province.

Unfortunately -- for the animals -- that is not the way this Biblical metaphor has been developed. While the relation between God and His special children, i.e., humans, was interpreted using the model of the loving parent, the relation between us and the rest of creation was interpreted using the model of the medieval feudal despot. Christian theologians interpreted the granting of "dominion" not as a solemn responsibility to care for what God has created but as a license entitling us to treat nature as our domain, as having been created for our benefit, as a resource to use as we see fit to satisfy our needs and desires. As a result, the idea that humans might owe respect to or consideration for animals became as unthinkable in Christian moral tradition as the idea that a feudal king was obligated to respect his serfs.

In the centuries following the death of Christ, Christian theologians turned to Greek philosophy to interpret Christianity in a way which would make sense to the intellectual community of Europe. Especially in the later Middle Ages, it was the philosophy of Aristotle that was thus employed. Aristotle declared that all things were governed and understood by four factors: the material of which they were made, their form or organizing structure, their maker, and their purpose. Applying this approach to the study of nature gave rise to the famous dictum that "Nature does nothing in vain," that is, everything in nature serves a purpose. Indeed, Aristotle organized all of nature -- which unlike the Christian view, included humans, i.e., "rational animals" -- into a hierarchical order in which the lower orders were there for the purpose of serving the higher orders. Aristotle's hierarchy was one of complexity, with the least complicated things at the bottom, e.g., mud and rocks, and the most complicated, viz., intelligent Greek men, at the top. Aristotle's "scientific" ordering of nature thus reinforced the Christian belief that all of nature existed to serve human ends and left animals still bereft of any sort of direct moral status or protection.

The best thing that the animals could do under this regime was enter the arena of moral concern under two indirect headings. First, under the standard moral and legal statutes concerning property, animals who were owned were protected against being harmed by anyone but their owners. Even some wild animals attained some protection in this way, since they were considered the property of the king, duke, or other local royalty. Second, all animals enjoyed the protection of being occasions for moral education. One of the most popular arguments for the humane treatment of animals was put forward by St. Thomas Aquinas in the thirteenth century. According to Thomas, we could not have any moral obligations directly to animals, since they lack souls; nevertheless, we should be considerate of the needs of animals and not be indifferent to their suffering, because if we get in the habit of being insensitive to them, we may well become insensitive to the needs and feelings of our fellow humans, which would be a morally pernicious development. Thus, although the animals were denied the full fruit of moral concern by the Christian-Aristotlean view of the world as a hierarchical, purposive, feudal order, animals were able to gather a few crumbs of respect and compassion as property and teaching devices.

It is worth noting that Aristotle did not confine his purposive, hierarchical ordering of things to inter-species relations. He extended it, logically enough, into our intra-species relations, contending that the less intelligent races were intended to be slaves for the more intelligent -- the Greeks being the most intelligent, of course. He also claimed that, for the same reason, women were intended to serve men. These claims again merely reinforced the sexism, anti-Semitism, and racism which have long infected Christianity, and did so emphatically during the Middle Ages. However, it was not these moral prejudices which led to the fall of the Christian-Aristotlean worldview of the modern era. Rather, it was the rejection by modern scientists of the category of purpose in understanding nature. People like Newton, Galileo, and other pioneers of modern science were able to devise ways of understanding and manipulating nature which did not involve presuming that anything in nature was created for a purpose. Rather, things were just the results of undirected causal forces, with the chain of causes stretching back ad infinitum. Since the concept of purpose had proven so unfortunate for the animals, we might expect that its demise would mark the beginning of a golden age for the animals, but, unhappily, that is not the case. The new science contended that except for the human mind--and perhaps not even with that exception--everything could be understood as a complex of machinery, nothing but gears, levers, nuts, and bolts. When applied to animals, this world metaphor led to the conclusion, made famous by the seventeenth century French philosopher and mathematician, Rene Descartes, that animals, like clocks, could feel neither pleasure or pain. They were merely "automata," said Descartes, God's ingenious robots. This conclusion removed animals even farther from the realm of moral concern than they had been under the Aristotlean rule, since the function of morality is precisely to protect and further the interests of those capable of feeling pleasure and pain. It is surely not an accident that the practice of vivisecting animals--nailing them to boards and then dissecting them while still alive -- was begun by the followers of Descartes at Port Royal.

Reaction against vivisection immediately followed its inception. The French philosophers Montaigne and Voltaire were particularly strident in rejecting the idea that animals were merely unfeeling machines who could be dissected with as little concern as one might have in taking apart a clock. Nevertheless, the seventeenth and eighteenth centuries, intellectually dominated by the successes of the physical sciences, very likely represent the low point for animals in mainstream, Western philosophy. However, in good dialectical fashion, it is from this moral desert that the fortunes of the animals will arise and progress in the nineteenth century.

STAGE II: "BE KIND TO ANIMALS"

Western moral thinking has been divided, roughly, into reflections about two families of ideas. The first of these we may call "the kindness family." It includes such ideas as benevolence, compassion, sympathy, charity, happiness, welfare, and friendship. Moral philosophies which focus on this family of ideas tend to express moral concepts in terms of seeking the good life, being altruistic or saintly, being a good friend or neighbor, pursuing the general welfare, and making the world a happier place in which to live. The moral philosophies of ancient Greece and nineteenth century Britain provide examples of such kindness-dominated morality. The other family of ideas we may call "the fairness family." It includes such ideas as justice, obligation, responsibility, rights, honesty, integrity, and commandments. Moral philosophies which focus on this family of ideas tend to express moral concern in terms of doing one's duty, fulfilling one's promises and other responsibilities, seeing that justice is done, and maintaining a clear conscience. The Old Testament and Puritanism provide examples of such fairness-dominated moralities. It follows that to enter the arena of moral concern in our culture is to be covered by at least one of these two families of ideas.

In the nineteenth century, animals finally got their hooves and paws through the door of kindness. Although Cartesian scientists may have been able to convince themselves that dogs screaming on the dissecting table were in the same category with clocks whose gears gnashed and whined when out of order, most people who came into contact with animals were too wise, or simply too honest, for that. And remember that in this era, when animals were still a primary source of transportation and the cities were not so insulated from the country, most people did still come into frequent, daily contact with animals. Having been freed of the limitations of the Christian dogma and the Aristotelan hierarchy, and rebelling against the vestiges of feudalism on many fronts, these people were free to acknowledge that feeling compassion for and being directly morally concerned about animals were neither heretical nor irrational. This breakthrough takes force with the founding of the first S.P.C.A.'s and other humane societies in nineteenth century England and the passage of the first significant humane legislation in the same era and locale.

The idea that animals are available for human service is, of course, not questioned here. The humane movement does not question the propriety of using animals for human transportation, food, clothing, or even science -- although this is also the era in which the first anti-vivisection societies are formed. However, the feudal view that those in power need not concern themselves with the needs and feelings of their inferiors is now displaced by the idea that we ought to be compassionate rulers who spare the animals we use and sacrifice any pain and suffering not necessary for that use or sacrifice. The model of the medieval despot has been replaced with that of the good shepherd who tends his/her flock not only for his/her benefit but also for theirs.

The idea of being kind to animals has grown and spread over the last century and a half until it now seems safe to say that it is the dominant idea in contemporary, Western moral thinking about animals. Victorian moralists touted humane concern for animals as one of the marks of a civilized society, and the first humane laws to protect dray animals have been developed into expansive codes prohibiting cruelty to animals and myriad public and private agencies devoted to protecting animals from abuse, protecting endangered species, rescuing animals in distress, and otherwise helping to relieve their suffering and ameliorate their condition. We spend a considerable amount of time, money, and energy caring for animals today, and we can be justifiably proud of living in an animal loving society.

Nevertheless, animals remain on the fringes of our moral concern today, not only in the sense that cruelty to animals is considered a minor crime but also in the sense that animals remain, like poor relatives, barely inside the door, with hat in hand. While our culture is committed to being kind to animals, that kindness has to compete with others of our concerns, such as those for abundant, inexpensive animal food products, for the freedom to do what we please with our property, and the best possible chance of having our ills cured and our lives extended. In this competition, our commitment to the humane treatment of animals often runs a poor second, third, fourth, or worse. For example, in order to spare farm animals the pain and stress caused by modern intensive farming techniques, we have not modified these techniques; rather, we cut the beaks off tightly caged chickens to stop them from pecking each other in frustration, and we keep veal calves in dim light in the belief that a drowsy calf is a contented calf. Like other recipients of charity in our culture, but even more so, animals do benefit directly from our sympathy for their plight, but they must make do with what is left over once our other more pressing concerns have been satisfied. Growing dissatisfaction -- among humans -- with this situation has led to the birth of "a new ethic for our treatment of animals."

STAGE III: ANIMAL LIBERATION

During the past fifteen years, our humane ethic has come under increasing, sharp criticism. "Kindness is not enough!" might be the slogan for this new group of animal activists. In terms of the analytical framework we have been using here, what is now being sought for animals is that they be covered not only by the kindness family but also by the fairness family. The situation today concerning animals is analogous to that two hundred years ago concerning slaves. While eighteenth century society felt comfortable with requiring only that slave owners treat their slaves compassionately, a small but growing group of people were demanding the abolition of slavery altogether. They contended that even if one was kind to his/her slaves, slavery was still an unjust institution in which the interests of one group were routinely sacrificed to fulfill the interests of another group. The slaves bore all the burdens, while the masters reaped all the benefits of slave labor, and that is the rankest sort of exploitation, no matter how benign the masters were to their slaves. And that is the way things remain with animals today.

Even where animal researchers adhere to and even exceed the requirements of the Animal Welfare Act to insure that their animals do not lack for veterinary care, anaesthesia, and painless death, these animals are still forced to acquire diseases, and to die for causes from which they will receive no benefit whatsoever. This is as intense an injustice as any humans have ever suffered, and given the magnitude of our exploitation of animals -- with several billion a year being raised and killed annually in the United States alone--this is a vastly more massive injustice than any humans have ever suffered.

Of course, most of our contemporaries still do not see our use of animals as being an injustice or as being covered by the fairness family of moral ideas at all. Although none of them would, in other areas, accept anything as long discredited as Aristotlean science and feudalism, when it comes to animals, they feel quite comfortable, thank you, with a hierarchical worldview that places them at the top and only marginally inhibits how they use their inferiors to satisfy their desires. In response to this self-serving moral complacency, philosophers such as Peter Singer, Bernard Rollin, myself, and others have been emphasizing that just as our basic moral concepts are color blind and sex blind, so they are species blind. For example, there is nothing in the logic of the Golden Rule to treat others as we would like to be treated by them which restricts it to people. Similarly, the altruistic ideal of setting aside selfish interests in order to do that which will be best for all concerned logically extends beyond the human family to cover all beings with interests. Again, pain, frustration, and boredom are evils because of how they feel, not because of who feels them; so our moral commitment to minimizing the misery of this world logically covers all those who can experience such evils, not merely those sufferers who happen to have Homo Sapiens genes.

Opponents of animal liberation often try to portray it as a product of implausible Eastern religions, such as that of the Jains, or of mysterious, probably empty constructions, such as "natural rights" and "inherent value." But that's simply not true. The liberating of animals from human exploitation is merely the next logical step in the progress of our everyday, Western moral concepts. Aristotle was the first major philosopher to say that slavery was morally pernicious--but his vision was ethnically and sexually limited: he objected only to the enslaving of Greek men; the enslaving of Persians and Egyptians and of women of all races he considered natural. It has taken us over two thousand years, and overcoming all varieties of religious and racial, as well as ethnic and sexual, prejudices to bring Aristotle's insight to its present, humanistic state of development on the idea that no people should be slaves. Conceptually, the basic shift here has been made in the past two hundred years, with the shift, at least as regards relations among people, from the idea that there is a natural hierarchy, with one group destined to serve the interests of another, to the idea that we should all be given an equal chance at a decent life and protection against being exploited by those stronger than ourselves. Liberating animals is nothing more than applying this same, thoroughly ordinary moral concern to those who differ from us not only in color, language, religion, and sex but in species.

Thus, liberating animals is not only the bringing of animals directly and fully into the arena of our moral concern, it is also the next logical step in our overcoming of our feudal heritage by substituting egalitarian for hierarchical presumptions. As this is accomplished, the same thing will happen in moral philosophy that has already happened in biology: the evolution of our concept of humanity, and we will come to recognize that together we form one living, morally significant and worthy community of interests on this planet.

The above piece was originally a lecture delivered at the "Forum '87" conference sponsored by the Animal Protection Institute and held in Sacramento, California, October, 1987. It was then published in "Between the Species," in 1988, and reprinted as a pamphlet by The Albert Schweitzer Center in 1989.

Steve F. Sapontzis is Professor of Philosophy at the California State University, Hayward, California. He is the author of the book Morals, Reason, and Animals, Philadelphia: Temple University Press, 1987. Interested readers can find a more extensive discussion of points raised in the article in that book.

The Latham Letter, Spring, 1990, Vol. XI, No. 2, pp. 5-10.

Living Miracles

The Reverend Dr. Andrew Linzey

I witnessed a miracle. The date was the 15th of May, 1988. The time was 9:39 in the morning. The place was my house--the Chaplain's residence--in Wivenhoe. I saw an immaculately formed, miniature human being pushed out of a mother's womb and into the world. Within a few moments, I held in my arms new life. Little things I still remember: a little hand clasping my finger; the wrinkled forehead; the small feet and tiny toes; the brilliant sun outside--and most of all an overwhelming sense of relief and thanksgiving. It was as though God in person had visited us. It was the birth of our fourth child, Jacob Peter.

Life is a miracle. That we are here at all is miraculous. Very shortly after Jacob was born, our three other children: Adam, Clair and Rebecca, came excitedly upstairs to greet the new child. I shall never forget their looks of wonder and amazement as they peered over the swaddling clothes and said, "Hello," to the new face in their midst. Life is grace, gift, miracle. We do not own it; it does not belong to us. In the face of it, we do well to stand in awe and wonder, gazing at the graciousness of the mystery we encounter, It was D.H. Lawrence who spoke of the need for "the sixth sense of wonder."

"Why who makes much of a miracle?" wrote Walt Whitman

"As to me I know nothing else but miracles, (whether they be) animals feeding in the fields,

"Or, birds, or the wonderfulness of insects in the air,

"Or the wonderfulness of the sundown, or of stars shining so quiet and bright,

"Or the exquisite delicate thin curve of the new moon in spring;

"These with the rest, one and all, are to me miracles."

God the Creator is the source of all these living miracles. Not just human life, but all life. The living world does not exist simply for us. It is not here simply for human use and pleasure. The miraculous world of living creatures exists because God loves them, and sustains them, and rejoices in them. If we do not hear this divine rejoicing throughout creation, then we shall live mean, narrow, self-centered lives. Three attitudes we need above all else: celebration, responsibility, and reverence.

First, then, celebration. Now some of you may say, "What do we have to celebrate when animals are used cruelly in factory farms and laboratories; when animals are made captive for human entertainment; and, when animals are wantonly slaughtered?" And I agree; the record of human behavior towards animals is worse than dismal. But one of the reasons why it is so dismal is precisely because we have so little humanity in us. We do not know how to celebrate, rejoice, and give thanks for the beautiful world God has made. If we treat it as trash it is because so many of us still imagine the world as just that. For too long Christian churches have colluded in a doctrine that the earth is half-evil, or unworthy, or--most ludicrous of all--"unspiritual." To help expand our consciousness, we should reflect upon these words written of St. Francis by an early biographer:

He rejoiced in all the works of the hands of the Lord and saw behind all things pleasant to behold their life-giving reason and cause. In beautiful things, he saw Beauty itself, all things were to him good. "He who made us is the best," they cried out to him. Though his footsteps impressed upon all things, he followed the Beloved everywhere; he made for himself from all things a ladder from which to come even to his

" As to me I know
nothing else but miracles,
(whether) they be animals
feeding in the fields,
   "Or birds, or the wonderfulness
of insects in the air..."

throne. He embraced all things with unheard of devotion, speaking to them of the Lord and admonishing them to praise him ... For that original goodness that will one day be all things and in all, was already shown forth in this saint as all things in all.

Second, in addition to an attitude of celebration, we need responsibility. Here we must not mince words. Dominion means responsible stewardship, but much of what we now do to animals is frankly tyrannous. We treat billions of animals every year simply as means to our ends, as renewable resources, as laboratory tools, as units of production, as commodities which can be bought or sold or dispensed with like worn out TV sets or empty cans of lager. Three to four million experiments on living animals every year, 500 million animals slaughtered for human consumption, 45 million laying hens in battery conditions, a thousand unwanted or abandoned dogs destroyed every week. Let no one think that our treatment of animals is a small matter. In terms of suffering alone it ranks as one of the most urgent moral problems confronting the human species. The time for piecemeal reform is long past. We need a revolution in moral attitudes towards animals.

I put it to you that it cannot be possible to honor and love the Creator of living miracles and also destroy these miracles wantonly as we do today. More than this: when we humans destroy life casually, recklessly, indifferently, we ourselves become unworthy of life. The line from Scripture I always find most pregnant of all is that from Genesis 6, when God reflected upon the human species and the violence they had brought into the world and declared that "I am sorry that I have made them." You may recall the end of that story: it was only Noah who had the good sense to realize that we are all in one boat together and who was saved.

Third, in addition to attitudes of celebration and responsibility, we need reverence. Moral exhortation, though desirable, is not enough. We shall not act responsibly if we have no reverence for life in our hearts. The Church needs to teach reverence for life as a major aspect of Christian ethics. So much of our ethical thinking is ludicrously bound with personal issues which leave to one side the whole business of actually caring for the world that God has made. So much of Christian ethics is pathetically narrow and absurdly individualistic. We deserve the admonition from St. Bonaventure:

Open your eyes, alert the ears of your spirit, open your lips and apply your heart so that in all creatures you may see, hear, praise, love and worship, glorify and honor your God.

One of the major problems with St. Francis, like to a lesser extent St. Bonaventure, is that the Church has not taken any practical notice of him. St. Francis preached a doctrine of self-renunciation, whereas the Church today remains overwhelmingly concerned with its own respectability. St. Francis lived a life of poverty, whereas the modern Church is as ever concerned about money. St. Francis, like Jesus, associated with the outcasts and the lepers, whereas the English Church consists predominantly of the middle classes. But the greatest challenge of all from St. Francis consists in this: Animals are God's creatures and because of that they are nothing less than our brothers and sisters. Unless we recognize that truth, and feel it in our hearts, the outlook for our own species must be grim. "We must turn back to what we have left of the capacity for wonder," writes Laurens van der Post, "only reverence for life can deliver us from our inhumanity, and from the cataclysm of violence awaiting us at the end of our present road."

I began with a story of birth. I want now to conclude by speaking of another birth. In his letter to the Roman Church, St. Paul describes creation as itself in a state of childbirth. "The creation itself will be set free from its bondage to decay and obtain the glorious liberty of the children of God." It is this picture of new birth, of world birth, that I want us to hold before us. According to the Christian reckoning of things, the world is going somewhere. It is not destined for eternal, endless suffering and pain. It has a destiny. Like us, it is not born to die eternally.

The fundamental thing to grasp is that we have responsibility to cooperate with God in the creation of a new world. And this is why sensitivity to suffering and abhorrence of cruelty is central to the Christian faith, indeed so central that a cruel Christian must be a contradiction in terms. "What is a charitable heart?" asks St. Isaac the Syrian, and he replies:

It is a heart which is burning with love for the whole creation, for men, for the birds, for the animals ... for all creatures. He who has such a heart cannot see, or call to mind, a creature without his eyes being filled with tears by reason of the immense compassion which seizes his heart; a heart which is softened and can no longer bear to see or learn from others of any suffering, even the smallest pain, being inflicted upon any creature. That is why such a man never ceases to pray also for the animals, for the enemies of truth, and for those who do him evil, so that they may be preserved and purified. He will pray even for the reptiles, moved by an infinite pity which reigns in the hearts of those who are becoming united with God.

I believe then that the Church must wake up to a new kind of ministry, not just to Christians or to human beings, but to the whole world of suffering creatures. It must be our human, Christian task to heal the suffering in the world. We must take every opportunity, following St. Francis, even in things that appear small and insignificant to lessen the burden of suffering upon the animals world.

If we are serious about following St. Francis, we shall want to take every opportunity to honor the Lord of life by refusing to be a part of wanton cruelty and death. In this way, we shall ourselves be reborn. I leave the final words to Albert Schweitzer who writes in his autobiography:

"I could not but feel with a sympathy full of regret all the pain that I saw around me, not only that of men, but that of the whole creation. From this community of suffering I have never tried to withdraw myself. It seemed to me a matter of course that we should all take our share of the burden of suffering which lies upon the world."

Copyrighted 1989, A. Linzey

The Reverend Dr. Andrew Linzey preached the above sermon at the Service for Animal Welfare and Thanksgiving for the Life of St. Francis at St. Edmundsbury Cathedral in July, 1989.

He is the Chaplain and Director of Studies, Centre for the Study of Theology in the University of Essex, and author of Christianity and the Rights of Animals, available through: Crossroad, New York.

The sermon is printed with the author's permission.

MISCELLANEOUS

Unexpected Teachers: A New Look at Our Pets

Karen Kaufman Milstein, Ph.D.

Historically and cross culturally, people have related to animals in a variety of fashions, often inconsistently. Animals have been seen as gods, slaves, workers and companions (Levinson). Within any given culture, the valuing and the treatment of animals is extremely variable, differing both from one individual animal to the next, and from one species to another. In our relationships with companion animals or pets, they are generally perceived as friends, though almost never as equals. Seeing eye dogs and other animals with which we have therapeutic (for us) relationships also serve as esteemed workers. Co-existing with our positive attitudes toward these companion animals is our ambivalence toward, maltreatment and abuse of other, generally non-pet species. This is only too apparent as we examine the very indifferent and often cruel treatment of lab and factory-farmed animals.

Shifting our focus now to planetary conditions, we see thoughtlessness and ruthlessness similar to what many animals and species must endure acted out in a global, environmental level. The 20th century has proven to be one of the most violent ever, with a major problem being our relentless, unceasing attempts to intensify production and increase consumption. We are finding, however, this greediness and shortsightedness cannot be continued without courting ecological disaster.

If we are to pull back from such impending disaster, one of the needed solutions may be to "make reparation" (Serpell) to the earth. This might include increasing our concerns for the animals, people, and the planet, and changing the underlying myths and values by which we live. To do so, we need to alter our perceptions and our philosophies of us vs. them, and man vs. nature. It is essential that we recognize that "together we form one living, morally significant and worth community of interest in this planet" (Sapontzis). Otherwise stated, whatever happens to the animal kingdom also happens to man (Knight). From both ethical and practical standpoints, it is incumbent upon us to act accordingly, rethinking our often automatic and frequently exploitive responses to the animals around us. As expressed by a long-ago Lakota chief, man's heart away from nature becomes hard: a lack of respect for growing, living things leads to a parallel lack of respect for humans (Nollman).

What do pets and our relationships with them have to do with the major challenges related to our views of nature and to planetary survival?, potentially, a good deal. Obviously, our companion animals are important to us, are valued. We see in our relationships with them the possibility that "human ascendancy is a phantom, an egotistical myth" (Serpell), and understand that animals are truly worthy of our caring. The risk remains, however, that we value only those animals we know personally and love as individuals, disregarding both other animals and the world itself from which all of us spring. Nevertheless, animals, as representations of nature which are brought into our immediate environment, can be a stimulus for a higher level of learning and for our personal growth.

Animals we care about can serve as our link with the earth and with natural processes to the extent that we are

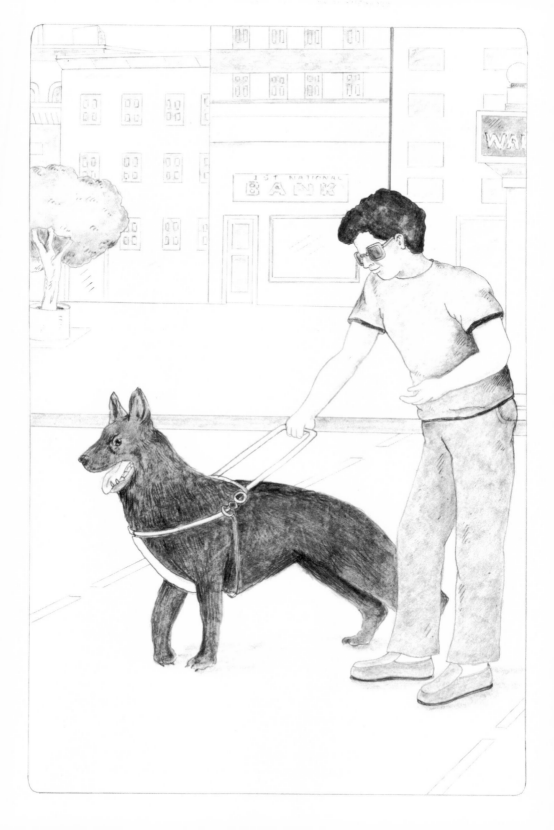

conscious and thoughtful in our relationships with them. If we treat our pets merely as "animated toys" (Katcher and Beck), without recognizing the reality of their own existence, then our current destructive perceptions continue, we do not see beyond ourselves, and we are likely to persist with our ruthless ravaging of our world and its inhabitants. If, on the other hand, we interact with our companion animals with real respect, acknowledging their differences from us even as we love them and live with them, then by extension we have the capacity to care about and value other species other than our own, both plant and animal. We become more capable of respecting, protecting and nurturing our planet. Through the sensitivity which this higher level responsibility for and love of our pets can develop in us, we may learn to relate to the natural world more intimately and responsibly. We benefit both through the healthier world which we then inhabit, and also through our own greater awareness and our ensuing development as spiritual beings. In order for this to happen, it is crucial that we be attentive, conscious and thoughtful about the significance of pet care, beyond the custodial responsibilities of food, water, and shelter which obviously we owe to our nonhuman friends. Always it is important to recognize their "other" reality and their connections with the natural world. Then we can learn more about our own nature and our intimate connections with our Earth, with our consciousness growing accordingly.

As well as being a potential stimulus for the development of more constructive attitudes and actions as described above, pet care offers us a wonderful opportunity to learn in another kind of way to care for our ultimate home, the natural world. In the concrete and practical arena of daily life with our pets, we engage in several kinds of transactions with our companion animals, through which they can help us to preserve our mental and physical equilibrium (Katcher and Beck). Foremost among these for our purposes is constancy. As we care for our pets, we share with them their daily cycles of feeding, excreting, playing, sleeping, etc., and we may also live through the life cycles of numerous pets, given their relatively short life spans. In this process, we experience some of the varying cycles of nature, and of constancy and continuity despite change, upheaval or loss (Katcher and Beck). We may appreciate more the ebbs and flows of our own experience, and the rhythms of life. Again, our appreciation and understanding of our natural world and of our place in it may be enhanced.

Thus, through our interactions with the nonhuman creatures in our lives, we may receive as much as or more than we give. We have the opportunity of becoming more sensitive, more perceptive, more thoughtful. A shift in consciousness can occur, bringing with it increased awareness and a resultant spiritual gain, as our connections with the planet and universe become more relevant and clear to us.

In summary, in their ofttimes delightful and sometimes trying fashion, our pets may invite and stimulate us to understand better ourselves, other species with which we share this planet, and our world itself. Our experience of the rhythms of life and of the interconnectedness of all its forms may be enhanced. For each of us, these learnings can be occasions of significant personal and spiritual growth. The necessary condition for this to occur is that we move beyond relating to our pets as cute and novel personifications of ourselves, existing primarily for our pleasure. Rather, we must come to know them as individuals with unique natures and needs. They can, as indigenous people have long known, truly be our guides to other worlds and realities, if we allow ourselves to be fully attentive and open to their essential natures.

Bibliography

Levinson, B.H. (1983). The Context for Companion Animal Studies. In A.H.Katcher and A.M. Beck (Eds.). New Perspectives on our Lives with Companion Animals. (pp 536-550). Philadelphia Press.

Serpell, J. In the Company of Animals, Oxford, England: Basil Blackwell, Ltd.

Katcher, A.H. (1983) Man and the Living Environment: An Excursion into Cyclical Time. In A.H. Katcher and A.M. Beck (eds). New Perspectives on our Lives with Companion Animals. (pp. 519-531). Philadelphia: University of Pennsylvania Press.

Sapontzis, S.F. (1990) the Evolution of Animals in Moral Philosophy. The Latham Letter, pp. 5-10.

Nollman, J. (1990) Spiritual Ecology. New York: Bantam Books.

Karen Kaufman Milstein, Ph.D. holds a Master's degree in human development. Dr. Milstein has many years experience working as a psychotherapist and resides with her family in the Philadelphia area.

The Latham Letter, Vol. XII, No.1, Winter 1990/01, pp. 4-5.

Webs

Stephan H. Johnsrud

Charlotte's Web by E.B. White remains a very popular classic and most people have read it. For those readers who are unfamiliar with it, Mr. Johnsrud's references will be confusing. Wilbur, a pig who was a companion animal for a little girl had to be relocated to a farm when he became too big. He arrived at his new home seriously depressed and lonely. Charlotte, a spider, introduced herself and the other farm animals to Wilbur and offered their friendship to him. When Wilbur learned of his impending fate, becoming bacon and hams, Charlotte spun a web incorporating the words "Some Pig!" It was perceived as a most unusual occurrence and the news of it caused Wilbur to become famous. It was necessary for Charlotte to spin three more webs to effect a permanent change for Wilbur's future. The result was that Wilbur would be able to live out his life in peace, security and friendship.

I identify with Wilbur in his pen striking up an unusual affinity with a much different species, a spider known as Charlotte. When Charlotte gave of herself in spinning the web declaring Wilbur "Some Pig!" things were never quite the same. A lady chaplain read that E.B. White classic, Charlotte's Web, to us in San Quentin Prison. I wondered about its meaning.

In 1983, I earned my Associate Arts degree from the College of Marin, Kentfield, California, thanks to the California Department of Corrections. I worked full-time in the prison education department in the daytime and attended classes at night. Programs were dwindling rapidly and I was very grateful for this opportunity. Receiving the diploma in the mail, I wondered what it meant.

December 15, 1983, having been transferred to the California Medical Facility at Vacaville, I turned in my first book recorded on tape for the Volunteers of Vacaville. A few weeks later, I became a full-time Reader and continue to this day. 4,666,200 recorded feet later, 833.33 hours per million feet. (I'll let you do the math.) A good part of seven years of my life have been spent in a tiny recording booth recording books-on-tape for the blind and visually handicapped. I am privileged to have my own anonymous Charlottes out there, spinning many interesting webs. One, for example, sent me a Textbook in Radiology. Great! I went absolutely wild on biology and life sciences earlier. She not only wanted it read, she wanted every single medical, chemical, physiological term s-p-e-l-l-e-d. A book full. OK. Can do. It just took awhile.

Another Charlotte is a seminarian in Southern California at what I'd call a rather Fundamentalist-type institution. (I'm either a Catholic or a Lutheran, I can't decide.) Now here something fascinating happens. Has anyone ever taken Koine Greek (New Testament) 101, 102, 103, in 1969 at Golden Valley Lutheran College, Lutheran Bible Institute ... and suddenly had to use it? I only earned C's and B's from Reverend Loddigs, but out came the awful declensions, "ho hay to, too tay too, ton tan ton," (phonetic). When word of this got around to other readers, suddenly I was the "Greek Expert." Going through a text in Systematic Theology, by a German theologian, well, by golly, that humble quarter of German at GV

[Golden Valley] also came in handy.

The next Charlotte might send The Mystical City of God, annotated version, by Mary of Agreda. Here, my Catholic experience and studies could be used, and as mystifying as this dated genre of writing seemed, I was no longer afraid of, and could sort of be one with, it as I read. In other words, keep my personal biases out of the text and not be a censor to my clients' beliefs and faiths.

Sure enough, the next style of spirituality might be from Billy Graham, the Evangelist, or Ellen White.

Whatever comes across our desk from clients, we divide up and read it. The Complete Works of Chuang Tzu, Modern Library Edition to The Complete works of St. John of the Cross, Institute of Carmelite Studies. 43 tracks, that is hours of listening in English, Latin, and the poetry in Spanish. Con mucho gusto!

And, of course, there have been many books in other fields, such as Federal Tax Manuals and Schedules. Every single chart! And there was a humongous law book, Agency and Principle, 2nd Edition, or was that vice versa? There were even two volumes of a U.S. Navy Diving Manual for a client with dyslexia; it had numerous oxygen-gas charts. Thank you, U.S. Navy for making texts of this kind familiar.

Some of these books have come to us because no one else in the entire reading-for-the-blind establishment has TIME. And it is not too difficult to imagine that people in prison have a lot of Time. Most of the time is completely squandered. Programs are almost non-existent, and would there be any motivation for them,.. even if they existed?

I feel like it is a very small pen and a very small web and feel almost selfishly grateful for this rare opportunity. I don't know if it can be expanded or duplicated or "catch on." I do notice that many outstanding volunteers in the community who have served as Braillists and transcribers for years are aging and do not seem to be replaced.

But let me return to a more subjective approach to my place in this human ecology. Some years ago I discovered I have manic-depression. After being hospitalized and prescribed Lithium, I returned to my reading booth, motor functions a bit clumsy, and discovered a book from the Latham Foundation: Dynamic Relationships in Practice. A photograph of a little girl in a wheelchair and holding a dog was on the cover. And I was sitting there, teary-eyed, reading it for some unknown person or persons who were blind. Those were some pretty big psychoanalytical words for a fellow fresh out of the hospital. Yet, we all managed to piece our various dysfunctions together into a marvelous web that worked. Later, I read an updated version, The Loving Bond. Once more, the idea of this unusual human ecology tugged at me. Navy instructors, nervous instructors (understandably) who came into San Quentin at night, all have said something to illuminate just one special page, perhaps, so maybe the young woman at the seminary can say to her professor, "luo tov duolon" (I loose the slaves).

We have appeared on the TODAY show, on other television programs and have the support of our warden, but for supporting us he really doesn't get the community credit he deserves. Twenty years later, we are sort of old news. We do have trouble finding readers who meet the literacy requirements. We have trouble keeping them, when we do find them. Well-intentioned people tour the facilities sometimes, even from other correctional facilities, and think it would be a fine idea. To date, their programs have not succeeded. The motto of the Volunteers of Vacaville is "Fiat Lux." Let there be Light.

I hope I have focused my light on the many Charlottes who need the books of their choice to be read for them, whether by inmates or free people at recording guilds on the streets. The nature of volunteerism in service to the blind is time-consuming and the new high-tech items are often not affordable for the average person.

And after all, someone read Charlotte's Web to me, a long time ago, and she did all right for someone with mild dyslexia. It's "Some Web!"

Stephan Johnsrud